Introduction

Contents

CW01594749

Overview

Construction site safety – Environment **is dedicated to the management of environmental issues related to the construction industry. It is part of the *Construction site safety* (GE 700) publication (see structure below), which covers health, safety and environment issues in the building and construction industry. It is designed to help managers, supervisors and small businesses understand how they should comply with, and put into practice, their legal, moral and social responsibilities.**

Construction site safety structure

Construction site safety is divided into the standard structure that is used across all core CITB publications.

Section A: Legal and management

Section B: Health and welfare

Section C: General safety

Section D: High risk activities

Section E: Environment

Section F: Specialist activities

Within *Construction site safety* each section is contained within a separate book, which has been designed to provide simple navigation for the user.

These sections are supported by **Section G: Checklists and forms**, available on the companion website. Section G contains interactive PDFs of all the checklists and forms that users will find useful on a daily basis. The forms and checklists follow the same structure as in Sections A to F.

Construction site safety – Environment

This publication provides practical, accurate and authoritative advice on all the important areas of the environment within the existing legal framework and industry requirements.

It explains what the law requires, what needs to be done and practical advice on how to implement and maintain the controls needed to manage the basics on site, as well as the more challenging elements. It will enable you to assess your legal responsibilities and those of others and decide how best to organise work activities in a safe and healthy manner.

This publication is also the official reference material for the *Site environmental awareness training scheme* (SEATS), a one-day course for supervisors and managers.

How to navigate

There are different ways to find the information you require within the sections. You can use the:

- ☑ initial contents pages at the start of the section
- ☑ more detailed chapter contents list
- ☑ index at the end of the section.

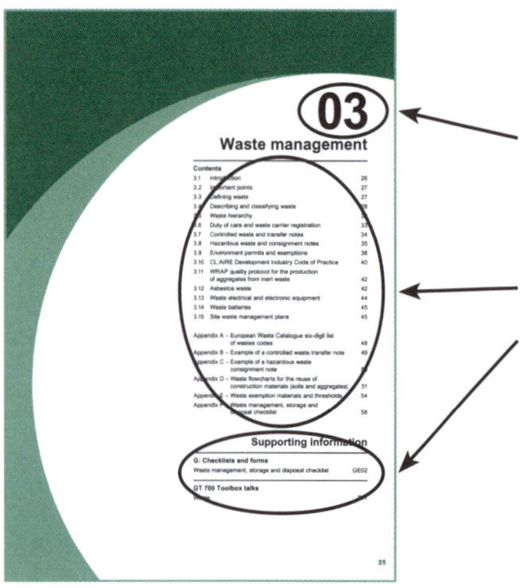

Chapter numbers have been included to help you find your way around the sections (for example, chapter one of Section A is 01, chapter three of Section E is 03 (pictured left), and so on). However, references to chapters within other sections are alpha-numeric (for example, A01, E03, and so on).

Each chapter contains a contents list at the beginning.

The chapter contents lists will, where relevant, also contain references to any Section G: *Checklists and forms* (available on the companion website) or GT 700 *Toolbox talks*, that support the chapter topic and that you may find a useful source of further reference.

Contents

Use of icons

A set of icons emphasises important points within the text and also directs readers to further information. The icons are explained below.

- Website/further info
- Example
- Question
- Ideas
- Notes
- Favourite
- Important
- Good practice
- Poor practice
- Caution
- Consultation
- Guidance
- Case study
- Quote
- Definition
- Interactive checklists and forms
- Video
- Shopping basket

Companion website

The companion website is a free resource that provides updates to the current GE 700 publication, to keep it up to date. It also supports the reader in progressing from the GE 700 book content to additional information available on the internet. Instead of having to type long website details into browsers, or searching on appropriate terms, the reader is directed to the companion website where all other sites of interest can be accessed by quick links.

The companion website supports GE 700 and it contains up to date information on:

- any amendments or updates to the current edition
- news (such as legislation changes, industry guidance and good practice)
- web-links, phone numbers and addresses
- interactive PDF checklists and forms
- the current edition of each section (book).

This icon indicates that further information (such as useful websites and links) can be found on the companion website at citb.co.uk/GE700companion

The companion website is regularly updated to ensure that the information is current.

Save the companion website address to your favourites, so it is always available when you need it.

Steps to access a link

The content on the companion website is structured in the same way as the current edition of GE 700.

To access information from a link in a book use the following steps to navigate the structure.

The example provided below is for the **Low Carbon Routemap and action plan** referenced in **Chapter E04 – 4.1 Introduction**.

Step 1	Open the companion website	citb.co.uk/publications/companion-websites/ge-700-companion/weblinks/
Step 2	Open the relevant section for the content required	**E: Environment**
Step 3	Open the relevant chapter	**E04 Energy management**
Step 4	Select the relevant link	**4.1 Low Carbon Routemap and action plan**
Step 5	Access the 3rd party referenced site	Green Construction Board site opens

Interactive checklists and forms

Interactive checklists and forms are available to download from the GE 700 companion website (identified by this icon).

To access an interactive checklist or form use the following steps to navigate the structure.

The example provided is for the statutory nuisance checklist, referenced in **E06 Statutory nuisance**.

Step 1	Open the companion website	citb.co.uk/publications/companion-websites/ge-700-companion/checklists-and-forms/
Step 2	Open the relevant section for the content required	**GE: Environment**
Step 3	Access the interactive PDF	Download the **GE06: Statutory nuisance checklist**

You are able to complete individual or multiple forms on your computer before saving or printing them to use, as required. The forms are compatible with both PC and Mac operating systems. You can also print off blank copies of the forms and fill them in manually, if required.

The checklists and forms are user-friendly and quick to complete, making the recording of important information a simple process.

Augmented reality

Augmented reality (AR) technology is used in our publications to provide readers with additional digital content such as videos, images and web-links.

This technology has been used to connect you with extra and complementary content. This can be accessed via your mobile device using the Layar app.

How to install and use the Layar app

- Go to the appropriate app store (Apple or Android) and download the Layar app (free of charge) to your mobile device.
- Look out for the AR logo (right), which indicates a Layar-friendly page or image.
- Open the Layar app and scan the Layar-friendly page or image (ensure that you have the whole page or the specific image in view whilst scanning).
- Wait for the page to activate on your device.
- Select one of the buttons that have appeared to access the additional content.

Where can I find augmented reality in this publication?

The table below identifies the pages in this publication that are compatible with augmented reality and the information that can be accessed when an active page is scanned.

GE 700 section	Location	Content
E[s]: Environment	Cover	Visit the GE 700 companion website
		Access the Section G: Checklists and forms
		Buy related products

01 Sustainable construction and the environment

Contents

E
01

Overview

The construction industry can affect the environment in a number of ways. It therefore has a major role to play in protecting natural resources and ensuring that they are passed on to future generations, in good order, for their enjoyment.

This chapter gives a general introduction to the environment. It will explain how the environment is defined, what local and global impacts put pressure on the environment and how it links with the overall concept of sustainable development.

This chapter also gives a general overview of the legal framework for the regulation of the environment and government environmental targets. It will also explain how the construction industry is responding to the environmental agenda by using fewer resources, less energy and promoting sustainable development.

1.1 Introduction

Rapid population growth and an ever-increasing demand for resources from economic development are placing huge pressures on our planet. In turn, this has led to an increase in all types of pollution and an acceleration of environmental damage. If everyone in the world lived as we do in Europe, we would need three planets to support us because we consume resources at a much faster rate than the planet can replenish them. People, consumption, production and the environment are all linked and have a major impact on each other. It is important, then, to consider how construction's contribution to future development can be achieved without causing any further damage. Sustainable living is about respecting the earth's environmental limits.

The terms *sustainability* and *sustainable development* were first established in the paper *Our common future*, released by the Brundtland Commission in 1987 for the World Commission on Environment and Development.

Sustainable development is the kind of development that meets the needs of the present generation without compromising the ability of future generations to meet their own needs.

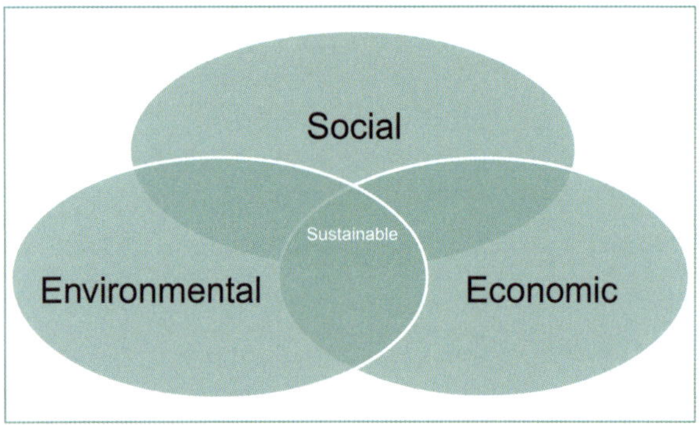

The main goal is to integrate the three most important pillars of sustainability that contribute to the achievement of sustainable development.

- ☑ **Environment.** Protection and enhancement of natural resources.
- ☑ **Society.** The wellbeing of people.
- ☑ **Economy.** Sustainable consumption and production.

Most people now recognise that economic activity cannot take place without considering the environmental impacts. For example, the extraction of aggregates for new development has a significant effect on the local environment, which must be taken into account.

The strategy of the UK Government for achieving sustainable development was first set out in its paper called *Securing the future – delivering UK sustainable development strategy*, which has four important priorities.

- ☑ **Sustainable consumption and production** (for example, through better product design, resource efficiency, waste reduction and sustainable procurement).
- ☑ **Climate change and energy** (for example, implementing measures to reduce carbon dioxide emissions).
- ☑ **Natural resource protection** (for example, promoting sustainable timber production and reducing the rate of biodiversity loss).
- ☑ **Sustainable communities** (for example, placing sustainable development at the heart of the planning system and enabling local communities to have a say in how they are run).

Progress in achieving sustainable development is measured against a set of indicators (currently 12 headline and 23 supplementary indicators, published in July 2013, further to consultation), which were originally outlined in *Securing the future strategy* where the impacts of production and consumption can have significant environmental impacts.

Current environmental indicators include greenhouse gas emissions, natural resource use, wildlife and bird population indices and water use.

www.

For further information refer to the *Securing the future: delivering UK sustainable development strategy* Government website.

1.2 Important points

The construction industry has a major role to ensure that the built environment is shaped in a way that delivers a sustainable future and reduces the environmental pressures on our planet from impacts such as climate change, resource depletion and biodiversity loss.

The UK and Northern Ireland construction industry is responding to this challenge through the Construction Leadership Council and Green Construction Board. It has, in partnership with Government, put in place action plans such as *Construction 2025: industrial strategy for construction*, which has set an aspiration for low carbon resource efficient construction, to deliver a 50% reduction in carbon emissions in the built environment by 2025.

For any construction project there will be a number of stakeholders who will have an interest in ensuring that the environmental requirements are being met: investors, clients, neighbours and regulators, to name a few. Effective environmental management will support your own legal compliance with environmental legislation and should ensure that all of these stakeholder requirements are met.

The environmental issues associated with construction are highly regulated. The UK and Northern Ireland Local Authorities and Environment Agencies have a key role in ensuring that applicable regulations are being complied with, and local impacts that can cause a nuisance (such as noise, dust and vibration) are minimised.

1.3 Sustainable construction

The construction industry has an important role to play in achieving these sustainable development priorities for the following reasons.

- ☑ The output of the construction industry is enormous; it is worth in excess of £100 billion per year.
- ☑ The construction industry accounts for around 6.5% of GDP, employs over 2 million people and supports 280,000 businesses.
- ☑ Buildings are responsible for almost 50% of UK carbon emissions, 50% of water consumption, about 30% of landfill waste and 25% of all raw materials used in the UK economy.

The construction industry has a significant opportunity to influence these issues and change the way we build. The Government's response to achieving sustainable development in construction was initiated through the strategy for sustainable construction, launched in June 2008. This joint industry and government strategy was based on a shared recognition of the need to deliver a radical change in the construction industry. Examples of industry's response to deliver sustainability in construction are highlighted below.

e.g. Government and construction industry response to improving sustainability in construction

- ☑ Establishment of the UK Green Building Council in 2008.
- ☑ Improvement in building energy performance through enhanced Building Regulations in 2010 and 2013 (Part L).
- ☑ The Waste Resources Action Programme (WRAP), operated between 2000 and 2015; halving waste to landfill (HWL), and the Built Environment Commitment for clients, contractors, designers and waste contractors.

  ***Note:** this knowledge base has now been transfered to the Construction Industry Research and Information Association (CIRIA).*
- ☑ All new buildings on the central government estate must achieve a Building Research Establishment's Environmental Assessment Method (BREEAM) rating of *excellent (refer to 1.9.1 for further details on BREEAM).*
- ☑ Establishment of the Green Construction Board in 2011, together with the low carbon routemap and action plan, which defines the work required for the built environment to meet the 80% reduction target by 2050. The *infrastructure carbon review* (ICR) is an important initiative launched by the Green Construction Board in November 2013 to demonstrate and encourage industry to embrace the link between lower carbon and lower cost across the supply chain. The Green Construction Board also produced a *Three years on report* in March 2015 and in May 2016 launched a new standard, developed by the Construction Leadership Council, publicly available specification (PAS) 2080, to establish a framework for the management of carbon by organisations working in infrastructure.
- ☑ Establishment of *Construction 2025* as part of the Government's industrial strategy, developed in close partnership with industry. It sets out a clear vision for UK construction, including *Sustainable – An industry that leads the world in low carbon and green construction exports*. The target is to achieve a 50% reduction in greenhouse gas emissions in the built environment by 2025.
- ☑ Government have relaunched the Strategic Forum for Construction, targeting payments, pre-qualification, procurement models and industry commitments.
- ☑ Government Construction Strategy 2016-20 committing to and identifying robust measurements and analysis of sustainability indicators.
- ☑ In April 2017 a new International Standard ISO 20400 (sustainable procurement guidance) was launched to help organisations develop and implement sustainable purchasing practices and policies with the supply chain.

- ☑ **For further information on ISO 20400 visit the ISO website.**
- ☑ **For further information on PAS 2080 visit the Green Construction Board website.**

1.4 Defining the environment

In its simplest terms, and according to the World Commission on Environment and Development, the ***environment*** is where we all live. More specifically, it can be defined as any physical surroundings consisting of air, water and land, natural resources, flora, fauna, humans and their interrelation.

1.5 Local and global environmental issues

Construction work can affect the environment in a number of ways, most noticeably at a local level. However, it can also contribute to wider regional or global issues. The three areas that could be affected are the atmosphere, water or land; the end result of any impact is usually described as pollution or contamination.

The use of energy, water and the depletion of natural resources also have an impact on the environment. Dust, noise and traffic are also a nuisance and can have a significant effect on the neighbourhood and local community.

1.5.1 Local environmental issues

The table below provides a summary of typical pollutants that cause local environmental impacts to atmosphere, land and water.

Atmosphere	Land	Water
■ Dust	■ Oils and fuels	■ Silt
■ Exhaust emissions	■ Chemicals	■ Chemicals
■ Gases or vapours	■ Lead	■ Concrete
■ Odours	■ Waste and litter	■ Contaminated water
■ Noise	■ Spillage of materials	■ Run-off
■ Light or visual amenity	■ Concrete	■ Effluent
■ Smoke	■ Asbestos	■ Oils and fuels
■ Radiation		■ Hazardous solid matter
■ Asbestos		■ Slurry

One solution to environmental pollution must not divert the problem to another area. For example, a solution to air pollution should not lead to water contamination.

1.5.2 Global environmental issues

1.5.2.1 Climate change

Certain gases in the atmosphere (principally carbon dioxide, methane, nitrous oxide and chlorofluorocarbons (CFCs)) form an insulating blanket around the planet. This allows the sun's rays through, but prevents some of the heat radiated back from the earth escaping, which warms the planet. This can be likened to the role of glass in a greenhouse, hence the term *greenhouse effect*. Climate change brings various risks for construction, the most obvious of these being higher temperatures and erratic weather patterns, including flooding. It is important to ensure that buildings are designed to adapt to these risks and to mitigate further impacts through energy efficiency and the use of renewable energy.

1.5.2.2 Acid rain

Acid rain is a collection of loosely related environmental problems involving acid substances and hydrocarbons. Burning fossil fuels, especially coal and oil (mainly in power stations and motor vehicles) produces acid gases (sulphur and nitrogen oxides). The presence of ozone in the lower atmosphere (harmful at this height), formed by the interaction of nitrogen oxides and hydrocarbons, also contributes to the effects of acid rain.

1.5.2.3 Deforestation

The world's forests are being rapidly depleted by logging, slash-and-burn agriculture and development projects (such as roads, mines and dams). In the 1950s, forests covered about a quarter of the earth's surface. At the turn of the 21st century, the figure is only one sixth.

Around 200,000 acres of rainforest are cut down or burned each day. That is an area equal to 233 football pitches lost every two minutes of every day, and 78 million acres lost every year.

Forests are essential for a healthy world. They play an important role in regulating the global climate by, for example, locking up large amounts of carbon dioxide during photosynthesis. Trees prevent erosion, flooding and the formation of deserts. Forests contain over half the world's plant and animal species and provide a home, fuel and food to many of the world's 200 million tribal people. They also provide the basis for most medicines and cures discovered to date. The construction industry uses a large amount of timber and is increasingly aware of the need to use this resource sustainably, hence why certification schemes (such as the Forestry Stewardship Council (FSC)) and the Programme for the Endorsement of Forest Certification (PEFC) have been established.

1.5.2.4 Biodiversity loss

The clearing of natural ecosystems (linked groups of self-sustaining plants), which is usually a result of agricultural, industrial development or a consequence of war, causes great local disruption to the natural environment. One major concern about the wholesale destruction of any natural habitat, particularly tropical forest, is that many species may become extinct as the ecosystem disappears. On a wider scale, migratory species that use a particular habitat during one season (such as the breeding season) might also become extinct.

When species lose their natural habitat to deforestation and other causes, they do not immediately disappear. Instead, they gradually die off over several generations, building up an extinction debt that must eventually be paid in full. The Brazilian Amazon has accrued a heavy vertebrate extinction debt, with more than 80% of extinctions expected from deforestation that has already taken place.

The destruction of habitats found in other ecosystems (such as wetlands, oceans and grasslands) has the same consequences as deforestation.

Irresponsible construction can have a devastating impact on biodiversity. In recognition, UK legislators have put planning processes in place to effectively manage this issue.

1.5.2.5 Resource depletion

The world's finite resources (such as coal, gas, oil, minerals and metals) are being rapidly depleted. These resources cannot be used in a sustainable way because they are not renewable. In reality, hydrocarbons (oil-based fuels and products) are finite resources because it takes millions of years for them to form: they cannot be replaced naturally within human timescales. Similarly, metals are also a finite resource; the amount of any metal available, as well as its chemical and physical usability, is reflected in its price. Forests are a resource that can be regrown, but they are currently being depleted *(refer to 1.5.2.3 Deforestation)*.

Large scale quarrying of finite resources

Water, especially clean drinking water, is being depleted and even in areas of Europe with regular rainfall this resource becomes limited after a short period of drought.

Construction is wholly dependent on the use of resources; without them, development could not happen. The drive for sustainable consumption and production will inevitably increase the costs for scarce resources and lead to technological improvements in the reuse and recycling of existing materials.

1.5.2.6 Ozone layer depletion

The ozone layer is a thin concentration of gas in the upper atmosphere. It protects the planet by filtering out harmful ultra-violet rays from the sun. These can cause skin cancer, eye cataracts, and restrict the growth of plants and other organisms. The layer was being depleted, especially above the North and South Poles, by artificial gases that destroy ozone molecules. The main offender was a family of chemicals called CFCs (chlorofluorocarbons), which were used in industry and household goods but have now been banned. Another group of destructive gases is called halons. These were used in fire extinguishers designed to fight electrical fires, especially in electricity sub-stations, computer rooms and aircraft. Worldwide action on ozone-depleting substances has significantly reduced this issue.

1.6 Environmental stakeholders

Environmental management terminology often refers to *stakeholders* who, in simple terms, are interested parties who have either a direct involvement in a construction project or who may be affected by its work. Typical environmental stakeholders for a construction project could include the following.

Stakeholder	Potential interest
Investors	Ensure that the project's sustainable credentials increase value.
Non-governmental organisations (NGOs) (for example, Greenpeace, Friends of the Earth and the Green Building Council)	Publicise any good (but usually bad) environmental practices.
Client or client's representative	Set the project environmental requirements.
Designer	Ensure that the client's environmental aspirations are reflected in the project design.
Planning authority	Approve the design, including relevant environmental requirements and set relevant planning conditions.
Local Authority	Ensure that the works comply with statutory requirements for local air pollution control, including noise and dust and issue relevant permits.
Environment regulators (for example, Environment Agencies and Local Authorities)	Act as a statutory consultee in the planning process and issue relevant permits and licences for the works (such as discharge consents).
Staff and contractors	Ensure that the project environmental requirements are met and lead by example.
Government funded support organisations (for example, WRAP and the Carbon Trust)	Provide support, advice and good practice and drive environmental improvement.
Residents and community representatives	Participate in the planning and consultation process and ensure that the works are carried out without causing nuisance from noise, dust, light and traffic congestion.

Early identification of all relevant stakeholders, together with regular communication and liaison, is important to avoid the likelihood of delays, legal intervention or causing a nuisance to local residents. Effective communication with the local community reduces the risk of complaint. Some typical methods of communication are shown below.

- ☑ Letter drops to local residents and businesses.
- ☑ Visiting owners or occupiers of sensitive homes and businesses.
- ☑ Attending local interest group meetings.
- ☑ Articles in local publications and newspapers.
- ☑ Displaying contact boards with relevant contact details so that interested parties can comment – this is a requirement of the Considerate Constructor Scheme. (*For further details refer to Chapter E06 Statutory nuisance.*)
- ☑ Establishing a telephone complaint line, email address or website if appropriate.

Identify all interested parties and regularly communicate with them on relevant environmental issues.

1.7 Regulatory framework for the environment

Construction sites have a number of legal obligations, many of which relate to protection of the environment.

Environmental policy and law is based on guiding principles that define the content of any rules and regulations. These principles are firmly embedded within legal frameworks.

- ☑ **Prevention** is better than cure, so policy measures are introduced to prevent environmental harm rather than remediate environmental problems once they have occurred.
- ☑ **Precaution.** Where there is uncertainty over the environmental science of an urgent problem, environmental protection must be prioritised rather than risk an impact occurring.
- ☑ **Polluter pays** principle requires that those causing (or potentially responsible for) environmental damage bear the financial costs of doing so. For example, this payment may be made through remediation of contamination, investment in pollution abatement technology and clean production plant, or through environmental taxation and charges.

Legal obligations do not only originate from the UK Government, or its devolved administrations. There are many sources of environmental law, starting at the international level, which are then transposed into UK law. In fact, most environmental law in the UK now comes from regulations made in Europe.

In the UK there are two general types of law.

- ☑ **Civil law**, which covers disputes between individuals or organisations.
- ☑ **Criminal law**, which deals with offences against the state.

Laws can be statutory (detailed by the state) or, in the case of common law, be based on the development of principles through case-law precedents where a basic law is interpreted under specific circumstances and the interpretation is used thereafter.

In the UK there are three levels of regulation relevant to criminal law.

- ☑ **Statutes or Acts of Parliament** are documents that set out legal rules and have normally been passed by both Houses of Parliament in the form of a Bill and have received royal assent. Acts of Parliament are also often called primary legislation. The Climate Change Act 2008 is a statute.
- ☑ **Delegated (or secondary) legislation** can take different forms.
  - Statutory instruments (SIs), through which the Secretary of State can issue specific regulations.
  - Orders issuing specific rules relating to a statute or part of a statute (for example, the Carbon Reduction Commitment Energy Efficiency Scheme Order issued under the Climate Change Act).
  - By-laws, which are made by the various tiers of local government, and can cover such matters as the establishment of internal drainage boards.
- ☑ **Guidance.** For some pieces of legislation the relevant Government department or regulator will issue guidance (such as those below) on the interpretation and implementation of the regulations.
  - **Statutory guidance**, typically in the form of a Code of Practice that sets out how to comply with the law.
  - **Non-statutory guidance**, typically a circular from a Government department, or technical guidance from either Government or the regulator (for example, DEFRA *Non-statutory guidance*).

1.7.1 Government policy-making departments

The main Government departments involved in developing and implementing environmental policy and legislation are outlined below. They work with the devolved administrations of Wales, Scotland and Northern Ireland.

Department for Environment, Food and Rural Affairs (DEFRA) is responsible for making policy and legislation, and works with others to deliver policies in the natural environment, biodiversity, plants and animals; sustainable development and the green economy; food, farming and fisheries; animal health and welfare; environmental protection and pollution control; and rural communities and issues.

Department for Business, Energy and Industrial Strategy (BEIS) is responsible for ensuring that the UK remains at the leading edge of strategies for industry, business, science, innovation, energy and climate change. Responsibilities include the development and delivery of a comprehensive industrial strategy and leading the government's relationship with business, whilst ensuring the country has secure energy supplies that are reliable, affordable and clean.

The Department for Business, Innovation and Skills (BIS) and the Department of Energy and Climate Change (DECC) have merged to form the Department for Business, Energy and Industrial Strategy (BEIS).

Department for Communities and Local Government (DCLG) is responsible for implementing planning policy, Building Regulations and building-related environmental standards (energy performance certificates and display energy certificates).

The Green Construction Board is a consultative forum for Government and the UK design, construction and property industry. It was established to ensure a sustained, high level conversation and to develop and implement a long-term strategic framework for the promotion of innovation and sustainable growth. The board owns and monitors the implementation, and will build on the Low Carbon Construction Action Plan and Routemap.

Construction Leadership Council (CLC) has been established to own and oversee industrial strategy for construction. Membership is composed of senior business people representing the main industry bodies and senior representatives of Government departments. The CLC will concentrate on achieving the following joint industry and government commitments from the construction strategy.

- 33% reduction in both the initial cost of construction and the whole life cost of assets.
- 50% reduction in the time from the outset to completion for new build and refurbished assets.
- 50% reduction in greenhouse gas emissions in the built environment.
- 50% reduction in the trade gap between total exports and total imports for construction products and materials.

www.

Further information on these departments can be found on the Government and Northern Ireland Department of Environment (DOE) websites.

1.7.2 Environment Agencies

- **England.** The Environment Agency was formed under The Environment Act of 1995 by merging The National Rivers Authority (NRA), Her Majesty's Inspectorate of Pollution (HMIP), the Waste Regulation Authorities and several smaller units from the former Department of the Environment (now DEFRA). At the same time, some of the functions of the Secretary of State were transferred to the agency making it one of the most powerful regulators in the world.
- **Northern Ireland.** Department of Agriculture, Environment and Rural Affairs.
- **Scotland.** The Scottish Environmental Protection Agency.
- **Wales.** Natural Resources Wales, linking with the Forestry Commission and Countryside Council for Wales.

The environment agencies exist to provide high quality environmental protection and improvement. This is achieved by an emphasis on prevention, education and vigorous enforcement wherever necessary. The overall aim of protecting and enhancing the whole environment is intended to contribute to the worldwide environmental goal of sustainable development.

The **enforcement powers** of the environment agencies are set out in their enforcement and sanctions statement and guidance, which outlines the circumstances under which they will normally prosecute. The enforcement powers available to the agencies will vary according to the nature and severity of the breach, but could include the following.

- A formal warning.
- A formal caution.
- Enforcement notices and works notices (where contravention can be prevented or needs to be remedied).
- Prohibition notices (where there is an imminent risk of serious environmental damage).
- Suspension or revocation of environmental permits and licences.
- Variation of permit conditions.
- Injunctions.
- Remedial works (where it carries out remedial works, it will seek to recover the full costs incurred from those responsible).
- Criminal sanctions, including fines, prosecution and imprisonment.
- Civil sanctions, including financial penalties, which were new powers available to the agency from January 2011.

The environment agencies can authorise officers to enter premises, including any land, vehicle, vessel or mobile plant. The occupier may need to be notified in advance and entry may only be permitted at reasonable times. Officers must abide by and not go beyond the instructions within the authorisation.

In the case of emergencies under Section 108 of the Environment Act persons authorised by the environment agencies have the authority to enter any premises, by force, with or without a warrant, and to:

- take any equipment or materials necessary as evidence
- make any examination or investigation
- direct that the premises are left undisturbed while being examined
- take samples and photographs
- instruct any relevant person to answer questions and to declare the truth of the answers given, and require the production of records.

The environment agencies can also carry out formal interviews under caution in accordance with the Police and Criminal Evidence (PACE) Act. The Environment Agency's *Offence response options* document sets out the options available to every offence that the Environment Agency regulates.

www.

For further information on Environment Agency sanctions and offences refer to the Government website.

1.7.3 Local Authorities

Local Authorities are responsible for various environmental pollution control functions, including those listed below.

- ☑ **Air quality.** Local Authorities are responsible for the management and assessment of local air quality, including the establishment of smoke control areas and the prohibition of dark smoke from chimneys under the Clean Air Act.
- ☑ **Contaminated land.** Under the Contaminated Land Regime, Local Authorities have a duty to inspect their land, to formally classify it as contaminated land, and require it to be cleaned up by the owner/occupier. Land can be designated as a special site under certain criteria (for example, land seriously affecting drinking water, owned or occupied by the Ministry of Defence or used for certain industrial activity). In these circumstances the land will be regulated by the environment agencies.
- ☑ **Local Authority Air Pollution Control (LAAPC).** For less polluting processes requiring an environmental permit (in England and Wales) or authorisation (in Scotland), the control of emissions to air alone is exercised by Local Authorities. Local Authorities have a duty to give prior written authorisation for processes under their control. This includes, for example, permits for mobile crushing and screening equipment and concrete batching plants.
- ☑ **Nuisance.** Complaints of statutory nuisance (such as noise, dust and odour) are dealt with by Local Authorities. They may serve an abatement notice where they are satisfied that a statutory nuisance exists or is deemed likely to occur or recur. *(For further information refer to Chapter E06 Statutory nuisance.)*
- ☑ **Planning.** Local Authorities have the responsibility of implementing and regulating national planning policy, including environmental impact assessments, tree preservation orders (TPOs) and authorisations for hedge removal. In many cases other regulators will be a statutory consultee, as part of the planning process.
- ☑ **Waste controls.** Local Authorities have some powers under waste legislation to stop and search waste carriers and confiscate vehicles suspected of waste crime. *(For further information refer to Chapter E03 Waste management.)*
- ☑ **Water companies.** The composition and quantity of industrial discharges to sewers are controlled primarily by the regional water companies (frequently referred to as sewerage undertakers). *(For further information refer to discharge consents in Chapter E05 Water management.)*
- ☑ **Consents to work near water.** On 6 April 2012, when a further phase of the Flood and Water Management Act was implemented, responsibility for regulating work on ordinary watercourses in most areas of England and Wales transferred from the Environment Agency to lead local flood authorities. Lead local flood authorities are unitary authorities where they exist and county councils elsewhere. *(For further information on water abstraction refer to Chapter E05 Water management.)*

1.7.4 Other regulatory organisations

There are a number of other regulatory bodies (shown below) that have specific regulatory and advisory responsibilities that are relevant to construction sites.

- ☑ **Canal and River Trust.** British Waterways ceased to exist in England and Wales and in its place the Canal and River Trust was set up in July 2012 to care for 2,000 miles of historic waterways. In Scotland, British Waterways continues to exist as a public body caring for the canals, operating as Scottish Canals.
- ☑ **Cadw** is the historic environment service of the Welsh Assembly Government, having responsibility for designated archaeological and heritage sites in Wales.
- ☑ **Historic England** is responsible for protecting historic buildings, landscapes and archaeological sites.
- ☑ **Health and Safety Executive.** There is an overlap between environmental legislation and health and safety legislation, which is regulated by the Health and Safety Executive (HSE), including the COSHH Regulations and the Control of Major Accident Hazards (COMAH).
- ☑ **Internal Drainage Boards** have powers under the Land Drainage Act (as amended) to undertake works on any watercourse within their district other than a main river. A board's district is defined on a sealed map prepared by the Environment Agency and approved by the relevant ministry. In addition, boards can undertake works on watercourses outside their drainage district in order to benefit the district.
- ☑ **Natural England** is responsible for conservation of wildlife and geology, including sites of special, scientific interest (SSSIs) and prevention of damage to habitats.
- ☑ **Scottish Natural Heritage** is responsible for designated ecological sites, geological and geomorphological sites, and protected species.
- ☑ **Historic Environment Scotland** is an executive agency of the Scottish Government and is charged with safeguarding Scotland's historic environment and promoting its understanding and enjoyment.
- ☑ **Department for Communities** is responsible for the historic environment in Northern Ireland.

1.8 UK environmental targets

The UK and EU environmental targets are identified in a large number of strategies for each environmental policy area. The list below provides a summary of the main construction targets for each of these policy areas.

Policy area	Target(s)
Waste	To achieve a target of 50% for the reuse and recycling of waste materials (such as paper, metal and glass) by 2020; the target for non-hazardous construction and demolition waste is 70%. Figures for 2013–14 suggest that these targets have already been met, with more than 80% of waste now being recycled. The target for landfill set out in the strategy for sustainable construction was to halve waste sent to landfill by 2012. There is now an ambitious expectation that projects should be working towards zero waste to landfill in the future.
Energy	To reduce carbon emissions by 80% (from 1990 levels) by 2050, with intermediate targets of 34% by 2020 and 57% by 2032. The UK also has a target to provide 15% of all energy from renewable sources by 2020. Revisions to Part L of the Building Regulations aim to reduce carbon emissions further, with the ultimate goal of zero emissions. With the cancellation of zero carbon policy in July 2015, the EU nearly zero energy buildings 2020 target may initiate further improvements in building regulations for energy efficiency standards. *Construction 2025* has a target to achieve a 50% reduction in greenhouse gas emissions in the built environment by 2025. From April 2018 new regulations will be enacted in England and Wales for any privately rented properties to have a minimum energy performance rating of E and be validated by an energy performance certificate (EPC).
Water	Reduce the consumption of water through cost effective measures, to an average of 130 litres per person per day by 2030, or possibly even 120 litres per person per day depending on new technological developments and innovation. Water quality standards are also set out in the Water Framework Directive.
Air quality	Objectives and target values for the protection of human health and for the protection of vegetation and ecosystems are set out in the UK National Air Quality Strategy for England, Northern Ireland, Scotland and Wales. It specifies objectives and target values for a range of pollutants including benzene, carbon monoxide and particulates.
Biodiversity	The UK is a signatory to the Convention on Biological Diversity (CBD) and is committed to the new biodiversity goals and targets 'the Aichi targets' agreed in 2010 and set out in the strategic plan for biodiversity 2011-2020. The UK has put in place a set of indicators to measure its progress towards meeting these targets annually: *UK biodiversity indicators in your pocket*.

www.

For further information on *UK biodiversity indicators in your pocket* visit the Joint Nature Conservation Committee website.

1.9 Construction sustainability assessment tools

Building Research Establishment's Environmental Assessment Method (BREEAM) and the Civil Engineering Environmental Quality Assessment and Award Scheme (CEEQUAL) are tools used by the construction and civil engineering industry to improve the sustainability of projects. Clients are increasingly requiring these standards as part of the project obligations. Many of the requirements set out in these tools are directly linked to the objectives and targets set out in the Government's strategy for sustainable construction discussed above.

1.9.1 BREEAM

BREEAM is an environmental assessment method for buildings. It has certified over 550,000 buildings since it was first launched in 1990.

BREEAM methodology is to assess and certify (using licensed assessors) the sustainability credentials of a building at the following two stages of the project life cycle.

- ☑ Design and procurement.
- ☑ Post-construction.

For new buildings (for example, BREEAM new construction fully fitted out), it measures the performance in nine categories (shown below) of environmental criteria, with the relative weighting for each category shown.

1.	Management	12%
2.	Health and wellbeing	15%
3.	Energy	15%
4.	Transport	9%
5.	Water	7%
6.	Materials	13.5%
7.	Waste	8.5%
8.	Land use and ecology	10%
9.	Pollution	10%
	Total	**100%**
	Innovation (additional)	10%

The **management category** details the requirements for a number of site-related issues. An example is the **MAN 03: Construction site impacts**, which requires the monitoring and reporting of energy consumption, water consumption and transport of materials and waste. It requires the procurement of site timber for formwork, site hoardings and for temporary works in accordance with the Government's timber procurement policy. It also requires construction site environmental management systems to be in accordance with ISO 14001 or equivalent and to implement good practice pollution prevention practices and procedures.

A further aim of BREEAM is to support innovation within the construction industry. BREEAM does this by making additional credits available for the recognition of sustainability-related benefits or performance levels that are currently not recognised by standard BREEAM assessment issues and criteria. Awarding credits for innovation enables clients and design teams to boost their building's BREEAM performance and, in addition, helps to support the market for new technologies and innovative design or construction practices.

The overall score for the assessment will be the percentage of credits achieved in each of the nine categories, which are then weighted to provide an overall percentage score. The total score will determine the BREEAM rating as shown in the table below.

Assessment	Credits achieved
Outstanding	85%
Excellent	70%
Very good	55%
Good	45%
Pass	30%
Unclassified	< 30%

To ensure that performance against fundamental environmental issues is not overlooked in pursuit of a particular rating, BREEAM sets minimum standards of performance in important areas (such as energy, water and waste). It is important to bear in mind that these are minimum acceptable levels of performance and, in that respect, they should not necessarily be viewed as levels that are representative of good practice for a BREEAM rating level. To achieve a particular BREEAM rating, the minimum overall percentage score must be achieved and the minimum standards, applicable to that rating level, complied with.

The BREEAM ratings are validated by BRE through the submission of evidence, together with the relevant fees by the licensed assessor.

www. **For further information refer to the BREEAM website.**

1.9.2 Code for sustainable homes

The *Code for sustainable homes* (the 'Code') is an environmental assessment method for rating and certifying the performance of new homes in the United Kingdom. However, it has been withdraw in England as part of the Government's housing standards review. In the consultation document the Government states: 'The conclusion is that the Code has been successful in doing its job in terms of pointing the way forward. In light of this, the Government does not now see a need for levels or separate carbon and energy targets in the Code – carbon and energy targets should be set in Building Regulations as we move towards zero carbon homes'.

As a result of this, Local Authorities should no longer require the Code as a planning condition for new approvals. However, where there are existing contractual arrangements (for example, with registered social landlords under the affordable funding programme 2015–18) it will be possible to continue to register and certify against the Code.

It should be noted that the Code is still applicable in Wales, and Scotland have their own sustainable requirements covered in the Buildings (Scotland) Act.

For further information refer to the Communities and Local Government website.

1.9.3 CEEQUAL

CEEQUAL is an assessment scheme for civil engineering. It assesses how well project and contract teams have dealt with environmental and social issues in their work. If used during the design, construction or maintenance phases, the CEEQUAL assessment is likely to positively influence the project's or contract's environmental and social performance.

CEEQUAL became part of the BRE group in November 2015. The scheme now operates alongside BREEAM – bringing together the world's leading sustainability assessment methods for buildings, masterplanning and infrastructure.

The CEEQUAL assessment scheme is available in three forms.

- **CEEQUAL for UK and Ireland projects.** Applicable to all types of civil engineering, infrastructure, landscaping and public realm works. There are five types of award available under this form of the scheme.
- **CEEQUAL for International projects.** Based on the 'for UK and Ireland projects' this scheme is applicable to projects anywhere else in the world. There are five types of award available under this form of the scheme.
- **CEEQUAL for term contracts.** Specifically created for the assessment of civil engineering and public realm works that are undertaken through contracts over a number of years and in a wide geographical or operational area. There is only one award type available under term contracts.

A CEEQUAL assessment is a self-assessment process carried out by a trained CEEQUAL assessor who is usually a member of the project or contract team, rather than hired only for the assessment task. Assessors use the questions set in the CEEQUAL manuals and an online assessment tool provided by CEEQUAL to decide on and capture the scores their work deserves, and to log the evidence justifying those scores.

The CEEQUAL assessment manuals (Version 5) are laid out in nine sections, which have been weighted. More details on what each section covers are given in the *Scheme description and assessment process handbook*, which can be downloaded for free as part of the CEEQUAL manuals.

1.	Project/contract strategy (optional)
2.	Project/contract management
3.	People and communities
4.	Land use (above and below water) and landscape
5.	The historic environment
6.	Ecology and biodiversity
7.	Water environment (fresh and marine)
8.	Physical resources use and management
9.	Transport

A fee is charged to cover the cost of the verifier and the administration and progressive development of the CEEQUAL scheme. The fee is based on the civil engineering value of the project or contracted works or, if applying early in the process, on the client's or engineer's estimate.

For further information refer to the CEEQUAL website.

1.9.4 Leadership in Energy and Environmental Design

Launched by the US Green Building Council (GBC) in 1998, the Leadership in Energy and Environmental Design (LEED) standard has become widely used both within the US and around the world. In recent years, UK based client groups have begun to ask for LEED certification alongside BREEAM.

Around 1.85 million square feet of buildings around the world are LEED certified every day.

Like BREEAM, LEED is voluntary and it can be applied to any building type and any building life cycle phase. It promotes a whole-building approach to sustainability by recognising performance in important areas of energy and water efficiency, CO_2 emissions, indoor environmental quality and sustainable use of resources.

LEED credits are weighted differently, depending on their potential impact. The greatest weighting is placed on energy and atmosphere, with sustainable sites and materials and resources also receiving a high weighting.

A total of 100 base points are available, with six possible innovation in design and four regional priority points. There are four levels of achievement: certified (40-49), silver (50-59), gold (60-79) and platinum (80 and over).

There are five different rating systems to cover different types of project, including new construction, LEED for existing buildings, LEED for commercial interiors, LEED for retail, LEED for schools and LEED for core and shell. Most building types are included in one or more of these systems.

LEED differs from BREEAM in a number of areas, and contractors should review requirements fully. Some differences are shown below.

- There is no need for an accredited assessor, as the US GBC assesses applications (although an extra credit is available where an assessor is used).
- Design requirements are linked to the American ASHRAE standards whereas BREEAM relates to UK Building Regulations.
- Some credits are calculated using US specific outputs (such as US dollars saved for credits relating to energy).
- Regional priority credits can only be obtained in the US.

For further information refer to the US Green Building Council website.

1.9.5 RICS SKA rating online assessment tool

This tool allows property and construction professionals and SKA assessors to design, specify, rate and certify fit-out projects for environmental impact, using the SKA rating fit-out benchmark system. Use of the tool is free and open to all. Projects can be certified by qualified assessors for an additional fee.

For further information refer to the RICS website.

1.10 Sustainable use of materials, energy and water – resource efficiency

Materials, water and energy (including transport) are all forms of resource, where efficiency of use has a considerable influence on construction times, costs and environmental impact.

Construction projects that use materials efficiently will have lower construction times and lower costs. Clearly this will lead to greater competitiveness, more repeat business and greater customer satisfaction. It also reduces the amount of resources that are taken from the planet and the amount of waste that the planet receives via landfill from construction work.

Inefficient projects are costly, late, use excessive resources, produce too much waste, are bad for image and lead to reduced client satisfaction.

The Government's *Construction 2025* places a top priority on resource efficient, low carbon construction. To achieve this outcome the project leadership must lead by example and communicate the correct behaviours, at all levels of the project, to emphasise the importance of resource efficiency.

For further guidance on resource efficiency, energy management and water management refer to Chapters E02, E04 and E05 respectively.

E
01

02

Resource efficiency

Contents

Supporting information

G: Checklists and forms

Overview

The construction sector is the largest consumer of materials in the UK, and the largest producer of waste. More efficient use of materials would make a major contribution to reducing the environmental impacts of construction, including carbon emissions, landfill and the depletion of natural resources.

This chapter provides a brief introduction to the environmental and economic benefits of improved material efficiency and waste reduction. It provides guidance on the responsible sourcing of materials, chain of custody for timber and life cycle analysis.

The chapter also highlights the benefits of reusing and recycling materials, including the use of recycled aggregates.

2.1 Introduction

Construction uses huge amounts of natural resources and accounts for 25% of all raw materials used in the UK. Historically, construction has been an inherently inefficient process, arising from the bespoke nature of on-site construction. This not only wastes a lot of money, it also produces high levels of waste materials and causes excess material extraction to replace those materials that have been lost through inefficient use.

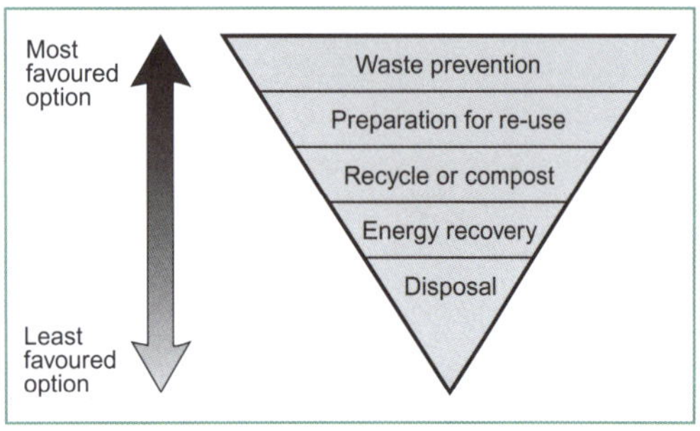

Waste hierarchy

The aim of modern construction is to move to the top of the waste hierarchy (waste prevention) and away from where it has been traditionally placed, which is at the bottom (disposal).

The Government responded to these challenges for sustainable construction by introducing the Site Waste Management Plans Regulations (England) (SWMP) in April 2008 and the Halving of Waste to Landfill by 2012 Commitment. The purpose was to drive improvements in resource efficiency throughout the construction process. The EU has also set out a roadmap for a resource-efficient Europe by 2020 (COM(2011) 571 final).

After consultation with industry, in July 2013, it was agreed that the Site Waste Management Plans Regulations (2008) would be repealed. *(For further information refer to Chapter E03 Waste management.)* However, some sites and contract requirements may still stipulate that a SWMP is used. The intention of a SWMP is to eliminate waste at the bid and design stage. It is a management systems approach that identifies where waste is likely to be produced in the construction process, and then seeks out ways to improve efficiencies and reduce waste.

Design for manufacture and assembly (DfMA) seeks to increase material efficiency by manufacturing modules in dedicated facilities that are then assembled on site. This has the potential to achieve zero waste, higher resource utilisation and to reduce the carbon impact.

2.2 Important points

Clients, designers, manufacturers, suppliers, contractors and installers have an important role in ensuring that the materials and products that form a construction project are both sourced and used with sustainability in mind. This is a pre-requisite of meeting certification to standards such as the Building Research Establishment's Environmental Assessment Method (BREEAM).

For further details on sustainable construction standards refer to Chapter E01 Sustainable construction and the environment.

Practical ways for improving resource efficiency and reducing waste

☑ Design.
- Specifying materials from sustainable sources; recycled and locally sourced.
- Designing the project to suit standard product sizes and to avoid site cutting.
- Designing for off site, pre-assembly and ensuring that products are responsibly sourced.
- Designing to allow a cut/fill balance and by utilising surplus materials in site features (such as landscaping).
- Looking for opportunities to source excess materials from other projects (for example, utilising the CL:AIRE register of materials and services).
- Specifying non-hazardous, low impact materials with low embodied carbon.
- Designing to allow for disassembly so components can be recovered and used elsewhere.

4

Holiday Inn

Waste duty of care

- Contain & Control
-
-
-

Waste transfer Notes — Keep 2 years
Consignment Notes — 3 years hazzard

(Section 34 of the Environ Act 1990)

What causes waste? / Late information / Poor stock control / Excessive packaging / careless measuring / Lack of planning / poor workmanship / Poor storage ......

Light risk → water surface polution not sewer

FWD — SW

Oil separator

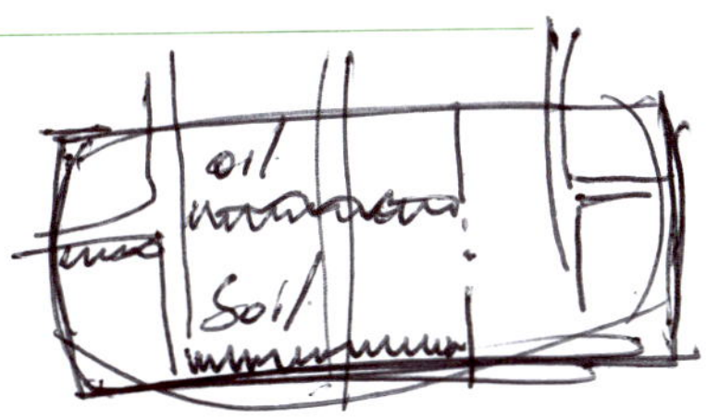

holidayinn.co.uk or 0800 40 50 60

E 02

al track record.

need to prove that
ducts used in their
uction products.
gic priority of low carbon
een construction

fficiency (net waste tool
ngineering projects. The
abling designing out

se visit the
vebsite.

rced. The aim is
nd encourage the
ng is also assessed
nvironmental Design –

nake the best
ation on the environmental
mental rankings are
kings that illustrate a

and then prove that their
scribes a framework for
the organisational governance, supply chain management and environmental and social aspects that must be addressed in order to ensure the responsible sourcing of construction products.

Independent, third-party assessment and certification against the requirements of BES 6001 then give the organisation the ability to prove that an effective system for ensuring responsible sourcing exists and adds credibility to any claims made. BES 6001 is also aligned to BREEAM so that products from manufacturers certified against this standard are able to score points against the responsible sourcing credits.

BS 8902 provides a framework for the development of sector certification schemes for responsible sourcing of construction products. It gives requirements for the management, development, content and operation of sector certification schemes for responsible sourcing and supply of construction products.

For manufacturers and suppliers, having certification to the environmental management systems (EMS) standard (ISO 14001) provides some evidence that the company has considered, and is reducing, the environmental impacts of their products. Relevant products assessed under the responsible sourcing elements of BREEAM are able to score points against credits where the certified EMS covers the main processes or main supply chain processes. For example, the responsible sourcing of bricks would need to demonstrate that the certified EMS covers manufacture (main process) and clay extraction (main supply chain process) to obtain maximum points.

2.3.1 Timber and chain of custody

Construction uses large amounts of timber, so it is important that timber is sourced from forests that are managed in a sustainable and ethical way. Illegal logging results in soil erosion, loss of biodiversity and uncontrolled development. It is therefore important to know where timber has been sourced from.

An estimated 78 million acres of forest are lost each year

The European Union Timber Regulation gives operators and traders an important role in ensuring that timber is sourced and supplied from legal and verifiable sources. This also applies to virgin wood products used for renewable heat installations (such as biomass boilers).

To combat illegal logging, which has serious environmental, economic and social consequences, the European Union adopted the Forest Law Enforcement, Governance and Trade (FLEGT) action plan in 2003. The action plan includes several measures that work together to stop the demand for, and reduce the supply of, illegal timber. Two important measures are voluntary partnership agreements (VPA) and the EU Timber Regulation. A VPA is a bilateral treaty between the European Union and a timber-producing country. Under this agreement, the timber-producing country agrees to control and license its timber exports as legal, and the EU agrees to accept only licensed imports from that country. These agreements are underpinned by strong systems for ensuring timber legality.

The EU Timber Regulation made it illegal, from 3 March 2013, to place illegally harvested timber and timber products on the EU market.

The legislation requires that due diligence is applied to all timber first placed on the EU market and also that traders, further down the supply chain, keep track of who timber or timber products were bought from and, where applicable, who they were sold to.

The regulation applies to two types of organisation within the EU timber supply chain.

- **Operators.** The bulk of the requirements apply to organisations who first place the timber products on the EU market.
- **Traders.** There are also requirements for traceability for all the other participants in the supply chain, prior to the final consumer sale. These organisations are all referred to as traders. (Construction companies, who procure timber from suppliers (either traders or operators) and supply timber to end customers, fulfil the role of a trader.)

As a trader, where you buy or sell timber products in the EU, you have to be able to identify the following.

- Where the timber products were bought.
- Where the timber or timber products have been sold, where applicable.

Chain of custody (COC) is a process that provides assurance about where timber has been sourced from. This is done by certifying timber from the forest to the final point of purchase and requires an effective audit process. The process tracks timber through each stage of the supply chain, from forest and logging, through sawmill, factory and distributor, to timber merchant and contractor. This provides a level of transparency and traceability to guarantee compliance with demands for ethically sourced timber products.

An example chain of custody is included in Appendix A.

There are more than 50 certification programmes addressing the many types of forests and tenures around the world.

Timber and chain of custody certificate number

In 2004 the UK Government established its own central policy for the procurement of timber – Central Point of Expertise on Timber (CPET). Whilst the CPET services were closed in March 2016 the policies they created are still applicable and available on the Government legislation website. Suppliers working for the Government are expected to comply with CPET requirements. The policies demand that all products derived from timber and wood must be from independently verifiable, legal and sustainable sources, or Forest Law Enforcement, Governance and Trade (FLEGT) licensed timber or equivalent sources.

These definitions have been revised to align with the EU Timber Regulations. The UK Government has approved the following two largest international forest certification programmes as timber certification schemes that meet CPET requirements.

- Forest Stewardship Council (FSC).
- Programme for the Endorsement of Forest Certification (PEFC).

The FSC and their global certification scheme allows consumers to identify, purchase and use timber and forest products produced from well managed forests.

For further information on the FSC and PEFC refer to the individual websites.

Schemes, such as BREEAM, have adopted CPET requirements for the purpose of demonstrating responsible sourcing of timber.

Some large construction projects have adopted a chain of custody certification scheme themselves as good practice.

EU Timber Regulations also cover virgin feedstocks of wood fuel for renewable heat installations (such as biomass) that are incentivised through the renewables obligation and renewable heat incentive (RHI) schemes. Non-domestic RHI participants will be able to demonstrate compliance with RHI sustainability criteria, by sourcing fuel only from a supplier registered on the approved RHI supplier list.

For further details on renewable heating energy refer to Chapter E04 Energy management.

For details of the biomass procurement requirements refer to the biomass supplier list (BSL) website.

2.3.2 Life cycle assessment (cradle to cradle)

Life cycle assessment (LCA) is described by the United Nations Environment Programme (UNEP):

> **life cycle thinking implies that everyone in the whole chain of a product's life cycle, from cradle to cradle, has a responsibility and a role to play, taking into account the relevant external effects. The impacts of all life cycle stages need to be considered comprehensively when making informed decisions on production and consumption patterns, policies and management strategies.**

A good LCA will identify where in the life of a product the main impacts occur and identify what can be done to reduce or mitigate these impacts. In construction LCA is applied to both the building itself and also to the products that go into that building.

The stages in the life of a building are material extraction, processing into a product, combination of products into the building, use of the building, end of life deconstruction and material recycling. Efficient recycling of materials at end of life has now given rise to the term *cradle to cradle*.

The Royal Academy of Engineering has identified that the typical costs for owning buildings are in the ratio of 1:5:200; one for construction costs, five for maintenance costs and 200 for building operating costs. Better investment decisions can be made by adopting whole-life costing and life cycle costing systems, which are vital to setting targets and to measuring and achieving long-term value and improved cost management.

Life cycle costing is a narrower assessment of the overall economic impacts of an asset, whereas the use of environmental costs in a whole-life analysis allows a true comparison between options, particularly where they are quoted as 'good' for the environment. For a major project (such as the construction of a nuclear power station) it is possible to calculate the environmental impact of making the concrete containment, the water required for refining the copper for the power plants and all the other components. Only by undertaking such an analysis is it possible to determine whether one solution carries a lower or higher environmental cost than another.

An important key construction material is cement and much work has been done on cement LCA. *The cement technology road map 2009* published by the International Energy Association (IEA) and the World Business Council for Sustainable Development (WBCSD) identifies the following action areas to reduce the LCA impacts of cement.

1.	Thermal and electric efficiency
2.	Alternative fuel use
3.	Clinker substitution
4.	Carbon capture and storage

PAS 2050, Specification for the assessment of the life cycle greenhouse gas emissions of goods and services, produced by the British Standards Institute, is a publicly available specification that provides a consistent method for assessing the life cycle greenhouse gas emissions of goods and services.

2.4 Reuse and recycling of materials

Maximising the reuse of materials on site can significantly reduce the amount of waste generated. For example, careful cut and fill analysis can ensure ground excavated from cuttings can be used as fill material elsewhere (such as within embankments) so that no waste is sent to landfill and there is no need to procure fill.

Not only can demolition materials be processed for aggregates and fill materials, designers should aim to use other materials (such as reclaimed bricks, timber and steel sections).

Other materials that may incorporate recycled content include plastics, aluminium, steel and steel reinforcement. Eco-reinforcement is a trademark for responsibly sourced reinforcing steel. It is a third-party certification scheme developed by the reinforcing steel industry to comply with BRE's *Framework standard for the responsible sourcing of construction products* (BES 6001).

Using construction materials that have been recycled and are low impact offers a number of clear environmental and business benefits.

- ☑ Demonstrating performance against corporate and sustainability policies.
- ☑ Reducing material costs (where locally reprocessed demolition materials are cheaper than virgin materials).
- ☑ Supporting sustainability goals, to meet the requirements of planning authorities.
- ☑ Providing a competitive edge through differentiation.
- ☑ Complementing other aspects of sustainable design.
- ☑ Conserving finite natural resources by reducing the demand for raw materials.
- ☑ Conserving energy and water, as recycled materials require less processing than extracting raw materials.
- ☑ Reducing air and water pollution, since manufacturing from recycled materials is generally a cleaner process and uses less energy.
- ☑ Reducing the amount of material that would otherwise go to landfill.

Taking a whole life cycle approach, designers should also take into account the possibility of reusing or recycling materials at the end of a product's life (for example, designing for dismantling and reuse of building components).

For a resource efficiency and responsible sourcing checklist refer to Appendix B.

2.4.1 Recycled aggregates

Aggregates include sand, gravel, crushed stone, slag, recycled concrete and, more recently, geosynthetic aggregates and recycled glass. The extraction of aggregates has a wide range of impacts upon the environment, including noise, vibration, vehicle emissions, visual impact and hydrogeology (study of the impacts on groundwater).

The largest volume of recycled material used as construction aggregate is presently blast furnace and steel furnace slag. At the end of the life of a building there is the potential to recycle aggregates and make them available as new construction materials. This is often the case where an existing site is being redeveloped.

There are many benefits of using recycled aggregates, including lowering embodied energy and reducing transport if produced on brownfield sites.

On-site processing of demolition materials into aggregates is classed as a waste operation and will require an environmental permit or exemption for the treatment and reuse of the material. The WRAP quality protocol of the production of aggregates from inert waste explains the requirements for ensuring that the processed materials meet end-of-waste status.

You have a responsibility to check that all relevant supplier documentation is correct to confirm that the protocol requirements have been met or that the relevant permit or exemptions that may apply are in place.

www.

- ☑ **For further information on the WRAP quality protocol refer to the WRAP website within the AggRegain section.**
- ☑ **The Construction Industry Research and Information Association (CIRIA) also have WRAP guidance on their resource efficiency knowledge base website.**

For further details on reducing waste and permitting requirements associated with the treatment of demolition waste and contaminated soils refer to Chapter E03 Waste management.

2.4.2 Reuse of excavated soils

Soil is vulnerable and is essentially a non-renewable resource. One hectare of topsoil, the most productive soil layer, can contain up to 5 tonnes of living organisms. Because it can take more than 500 years to form a 2 cm thickness it is, in practical terms, non-renewable.

Soil, if managed carefully, can be re-incorporated into the project, without the need to remove it from site because of poor design or contamination from other materials.

Re-incorporating soil into the project is an efficient use of resources

For further information on reusing soil from other projects refer to the CL:AIRE website.

The Department for Environment, Food and Rural Affairs (DEFRA), in support with WRAP, has prepared a construction *Code of Practice for the sustainable use of soils on construction sites*. It contains guidance to those involved at all stages of construction projects, from the developer, designer, contractor and sub-contractor (earthworks and landscape) to the regulator.

For further information on the *Code of Practice for the sustainable use of soils on construction sites* refer to the DEFRA website.

E 02

Appendix A – Example chain of custody certificate for timber

Certificate

Registration Certificate

Certificate Number TT-COC-001744

10 October 2003	10 October 2008	09 October 2013
Date of Initial Registration	Date of Issue	Date of Expiry

This is to certify that

Paterson Timber Ltd
140 Elliot Street
Glasgow
G3 8EX

meets the requirements for FSC Chain of Custody Standard FSC-STD-40-004 (version 2.0) within the BM TRADA Certification Limited Scheme for TRADA-Trak chain of custody certification.

The scope of the approval in respect of processes and products which are eligible to carry the FSC logo are as shown in the Schedule

This certificate cannot be presented as evidence that certified wood products have been supplied.

In all cases, such evidence must be conveyed on the corresponding sales invoice.

Signed on behalf of BM TRADA Certification Ltd
BM TRADA Certification Ltd, Stocking Lane, Hughenden Valley, High Wycombe, Buckinghamshire, HP14 4NR, England
FSC Trademark © 1996 Forest Stewardship Council A.C. FSC-SECR-0021

Registration schedule

Registration Schedule

Forest Products Chain of Custody Certification

Certificate Number BMT-COC-0015

Establishment	Address	Number
Paterson Timber Ltd	140 Elliott Street Glasgow Scotland G3 8EX	038/BMT-COC-0015

Scope

The Importation, Storage, Distribution and Sales of Independently Certified Timber and Timber Products

Products Covered by Certification

1. **Malaysian Timber Certification Council: "Requirements and Assessment Procedures for Chain of Custody Certification".**
2. **Palliseo/CIFM – REGISTRATION NUMBER OLB-CERTIFICATION – 050701.**
3. **IFO – STATEMENT NUMBER SGS-TLTV-0001.**
4. **SIFORCO – STATEMENT NUMBER SGS-TLTV/LP-0005.**
5. **Smartwood Verifiation of Legal Origin (VLO:) Verification code SW-VLO-0009.**

03 November 2004	02 November 2007	01 November 2012
Date of Initial Registration	Date of Issue	Date of Expiry

Appendix B – Resource efficiency and responsible sourcing checklist

Company name		Project title	
Location		Contract no.	

Resource efficiency and responsible sourcing

	Yes	No	N/A
1. Has the site considered alternative techniques for the efficient use of resources (such as off-site manufacture, design to fit modular sizes, manufacture to fit design and design for disassembly)?			
2. Has the design taken into account the reuse of reprocessed demolition materials or the use of recycled aggregate or excavated soil and materials as part of the site waste management plan?			
3. Are there procedures in place to identify and order the correct type and quantity of materials?			
4. Are there procedures in place to ensure the materials are sourced responsibly (such as only using timber grown under sustainable timber schemes and monitoring of supply chains to report and avoid manufacturers who use slave labour) and records maintained for traceability purposes?			
5. Are deliveries being inspected to avoid damage during unloading, deliveries to wrong area of the site or accepting incorrect specification or quantity?			
6. Are stockpiles and material storage areas away from traffic routes to prevent accidental damage?			
7. Are materials being stored to prevent repetitive handling?			
8. Are materials being stored with the appropriate security to prevent loss, theft or vandalism?			
9. Are stockpiled materials being stored to prevent cross-contamination with other materials or wastes?			
10. Are stockpiled materials being stored away from sensitive areas, drains or watercourses?			
11. Are materials being handled with the correct equipment to prevent accidental damage?			
12. Are hazardous/COSHH materials being stored, issued and disposed of correctly?			

Comments

Name		Position		Signature		Date	

For an interactive PDF version of this form visit the companion website.

E
02

03
Waste management

Contents

Supporting information

G: Checklists and forms

GT 700 Toolbox talks

Overview

The construction sector produces around 100 million tonnes of construction waste per annum, which is nearly a third of all waste produced in the UK.

This chapter gives a general overview of the legal framework for the regulation of waste. It will outline how waste is defined and classified for the purpose of disposal and will introduce the waste hierarchy.

This chapter also gives you an overview of the permits that are required for the management of waste and the documentation required under the duty of care for its safe and environmentally sound disposal. It will provide an overview of various types of waste, including hazardous waste, Waste Electrical and Electronic Equipment (WEEE) and waste batteries.

3.1 Introduction

Of the 100 million tonnes of construction waste produced by the construction sector per year, a proportion ends up in landfill sites. This waste has other environmental impacts (such as the generation of greenhouse gases) that contribute to climate change.

Waste is a by-product of the inefficient use of valuable resources that could be reused on site or recycled. The simple operation of cutting a brick in half, or sawing the end off a piece of wood, is producing waste if that half brick or off-cut is not reused.

Waste costs money to produce, in terms of the materials thrown away, and also money to dispose of, along with costs for the skip or lorry to remove it. Landfill tax rates will continue to rise to provide a financial incentive to reduce the amount of waste being sent to landfill.

Research carried out on housing projects has indicated that the true cost of disposing of the material is around ten times the cost of the skip. This includes the labour to fill the skip and the cost of purchasing the materials in the first place.

Typically skips account for 0.5% of the build cost. Overall waste and its disposal can amount to around 5% of the project costs. This is valuable lost profit for the project. Making the best use of materials and resources is therefore essential in reducing waste.

The incorrect or inappropriate disposal of waste (such as fly-tipping) is illegal, unsightly and can damage the environment for many years. To tackle the problem of waste crime Local Authorities have the power to stop and search vehicles thought to be involved in illegal waste activity.

The revised Waste Framework Directive has set a target of 70% (by weight) preparing for reuse, recycling and other recovery of construction and demolition waste by 2020.

Poor design, insufficient attention to the generation of waste during procurement, and poor control and supervision by site management, including improper or unsafe systems of work on site, increase the production of waste. Waste materials lying around on a construction site increase the risk of fire and have the potential to cause people to trip and injure themselves.

Legislation imposes conditions and obligations on how contractors in the construction industry may dispose of any waste produced during work both on and off site.

3.1.1 Waste regulation authorities

The principal waste regulation authorities are the Environment Agency (EA) in England, the Northern Ireland Environment Agency (NIEA) in Northern Ireland, the Scottish Environment Protection Agency (SEPA) in Scotland, and Natural Resources Wales (NRW) in Wales.

The environment agencies took over these functions from Local Authorities and their responsibilities include dealing with the application and enforcement of waste management licences, permits and exemptions, waste carriers' licences and the duty of care regime *(refer to 3.6)*.

Local Authorities are also responsible for dealing with local air pollution control (LAPC) matters related to waste (such as the issue of permits for crushing equipment).

The improper disposal of waste is illegal and can lead to prosecution and even imprisonment. The maximum penalty for waste crime is a £50,000 fine and/or imprisonment for six months.

3.2 Important points

- ☑ Producers of waste must correctly identify whether surplus materials are waste and classify it as non-hazardous or hazardous with reference to the European Waste Catalogue six-digit list of wastes codes.
- ☑ With reference to the standard industry classification (SIC) code, producers of waste must correctly identify the SIC and which area of construction the waste originates from (for example, construction of buildings, civil engineering or specialist construction activities).
- ☑ Producers of waste have a legal duty of care to ensure that it is passed on to an authorised person with the correct technical competence and holding a relevant environmental permit or licence.
- ☑ All contractors who carry or collect construction and demolition waste must have a waste carrier's licence.
- ☑ All waste transfers must be supported by the correct document (a controlled waste transfer note) for non-hazardous waste. The transfer of hazardous waste requires a consignment note. Both of these documents must include a declaration that the producer of the waste has considered the waste hierarchy in deciding to dispose of the material.
- ☑ In Wales, producers of 500 kg or more hazardous waste (such as oils or asbestos) must register their premises with the NRW (Wales). Failure to do so could result in a fine of up to £5,000.
- ☑ All waste treatment or disposal facilities should have an environmental permit (England and Wales) or waste management licence (Northern Ireland and Scotland) unless they have a registered exemption from the EA, NRW, NIEA or SEPA.
- ☑ Where materials are treated or processed on site before being suitable for putting back into the works then consideration must be given as to whether this activity requires an environmental permit or registered exemption. Compliance with schemes, including the Contaminated Land: Application In Real Environments (CL:AIRE) Code of Practice or the waste and resources action programme (WRAP) quality protocol for the production of aggregates from inert waste could avoid the requirement of an environmental permit for reuse of the processed material.

3.3 Defining waste

The starting point for classifying something as *waste* comes from the revised European Waste Framework Directive (WFD) (2008/98/EC). It states that:

> **... *waste* shall mean a substance or object which the holder discards, intends to discard or is required to discard.**

Waste is also defined in the Environmental Protection Act as:

> **... any substance which constitutes a scrap material, an effluent or other unwanted surplus arising from the application of any process or any substance or article, which requires to be disposed of, which has been broken, worn out, contaminated or otherwise spoiled.**

In practice, however, this definition has been tested by EU and UK case law; determining whether or not a substance is waste depends on applying the right legal tests. A substance can be classed as waste even if the producer still has a use for it or if other people are prepared to pay for it. This is important because whether or not a material is waste determines whether a complex body of legal rules and restrictions govern what can be done with it.

Surplus materials are generally not waste while they remain as the original manufactured product and do not need to be re-processed.

However, construction or demolition waste that has been generated as part of the works could be classified as waste until it has been processed (for example, crushed and screened) and recovered back into the permanent works. In these cases there will be various requirements to demonstrate that the material has achieved an **end of waste** status.

Materials that meet the end of waste test must satisfy the following criteria.

- ☑ The material must be converted into a distinct and marketable product.
- ☑ It can be used in the same way as an ordinary soil or aggregate.
- ☑ It can be used with no negative environmental effects.

Aggregates manufactured from construction and demolition waste complying with the WRAP quality protocol, for example, can normally demonstrate that the material has achieved an end of waste status. Likewise, complying with the CL:AIRE development industry Code of Practice will also help you to achieve end of waste status for the treatment and use of contaminated excavated materials.

(For further information on the CL:AIRE Code of Practice and WRAP quality protocol refer to 3.10 and 3.11 respectively.)

E
03

3.4 Describing and classifying waste

Wastes will always fall into one of three categories.

- ☑ **Never hazardous** (for example, clean bricks or glass).
- ☑ **Always hazardous** (for example, insulating materials containing asbestos).
- ☑ **May, or may not, be hazardous and need to be assessed** (for example, contaminated soils).

What makes a waste hazardous is whether it contains any hazardous substances above certain thresholds that make it display a certain hazardous property. Fifteen different hazardous properties exist, from HP1 to HP15, together with persistent organic pollutant (POP) criteria (for example, HP1 is explosive and HP6 is acute toxicity). The threshold concentration levels for the relevant hazardous properties are defined within the guidance on the classification and assessment of waste (WM3). Waste can be referred to as **active**, for hazardous waste, or **inactive**, for inert waste (such as rocks and bricks).

www.

For further information refer to the joint agency technical guidance *Waste classification – Guidance on the classification and assessment of waste* (WM3).

The most appropriate method of classifying waste, where it needs to be assessed, is to identify the hazardous constituents or chemicals in the waste, determine the hazard statement codes and hazardous properties of these substances and then to use their concentrations to identify whether they exceed the threshold levels of any of the hazardous properties. The safety documentation supplied with any product should provide sufficient information to make this assessment. For contaminated soils, however, detailed testing would need to be carried out by competent staff.

The **European Waste Catalogue (EWC)** is a standard six-digit coding system that describes and categorises different types of waste. These waste codes are arranged in 20 sections; Section 17 contains the codes for construction and demolition waste.

Appendix A includes the full Section 17 list of wastes codes relating to construction and demolition waste.

The list of wastes legislation in England and Northern Ireland has been revoked and reference is now made to the equivalent wastes in the EWC.

The list of wastes refers to hazardous and non-hazardous entries. Where an entry is marked with an asterisk it is classified as **hazardous waste** if it meets one of the following criteria.

- ☑ The hazardous entry makes no reference to hazardous substances (for example, 17 06 05* construction materials containing asbestos). These types of entries are always hazardous and are called **absolute hazardous entries (AH)**.
- ☑ The hazardous entry refers to a waste containing hazardous substances where the concentration levels of these hazardous substances exceed the threshold limits (for example, 17 05 03* soil and stones containing hazardous substances). These entries are called **mirror hazardous entries (MH)** as it depends on the concentration of hazardous substances to determine whether they are hazardous or not. Wastes containing hazardous substances below the threshold limits are non-hazardous.
- ☑ The lists of wastes also contain non-hazardous mirror entries (MN). The non-hazardous mirror usually (but not always) has a defined link to its mirror using the words 'other than those mentioned in....' (for example, 07-01-12 sludges from on-site effluent treatment other than those mentioned in 07-01-11 MN).

e.g.

Typical construction wastes from Section 17 of the list of wastes

- ☑ Bricks – 17 01 02 Non-hazardous.
- ☑ Concrete – 17 01 01 Non-hazardous.
- ☑ Wood – 17 02 01 Non-hazardous.
- ☑ Waste hydraulic oil – 13 01 13* Absolute hazardous.
- ☑ Plasterboard – 17 08 02 Non-hazardous (but must be segregated from other wastes).
- ☑ Insulation containing asbestos – 17 06 01* Mirror hazardous. (Will be hazardous if concentrations exceed thresholds.)
- ☑ Contaminated soil – 17 05 03* Mirror hazardous. (Will be hazardous if concentrations exceed thresholds.)
- ☑ Mixed canteen waste – 20 03 01 Non-hazardous.

The disposal of hazardous waste arising from construction operations, or from contaminated land, is dealt with under the Hazardous Waste Regulations (in England, Northern Ireland and Wales) or the Special Waste Regulations (in Scotland). *(For further information refer to 3.8.)*

All waste transfer documentation must include the relevant EWC six-digit code that describes and categorises the waste, and the relevant SIC code describing the area of construction from where the waste originated.

For the purpose of disposal of **waste to landfill** there are three classes of waste.

- **Inert waste** that will not decompose to produce greenhouse gases (such as rubble, concrete and glass).
- **Non-hazardous waste** that will rot and decompose (such as timber, food and paper) and does not contain hazardous substances.
- **Hazardous waste** that has substances in sufficient concentration to make it possess one or more of the 15 hazardous properties (such as explosive (HP1) or acute toxicity (HP6)) and is hazardous to human health or the environment (such as asbestos or oil).

There are strict criteria for the acceptance of waste at each of these three types of landfill site. Certain types of waste, particularly contaminated soil, would have to be tested in order to demonstrate that it meets the relevant waste acceptance criteria.

The landfill directive introduces a hierarchy of waste characterisation and testing known as the **waste acceptance procedures**. The three levels are shown below.

Level 1. Basic characterisation. A thorough determination, according to standardised analysis and behaviour-testing methods, of the leaching behaviour and/or characteristic properties of the waste.

Level 2. Compliance testing. A periodic testing of regularly arising wastes by simpler standardised analysis and behaviour-testing methods to determine whether a waste complies with permit conditions, and whether a waste with known properties has changed significantly. The tests focus on the main variables and behaviour identified by basic characterisation.

Level 3. On-site verification. This constitutes checking methods to confirm that a waste is the same as that which has been subjected to compliance testing and that which is described in the accompanying documents. It may merely consist of a visual inspection of a load of waste before and after unloading at the landfill site.

Before sending waste to landfill, waste producers and landfill operators must ensure that they know all of the properties of the waste, relevant to its potential for pollution or harm to health, and the options for the management of the waste.

There are certain types of waste that are banned from disposal to landfill (shown below). These must either be recovered, recycled or disposed of in other ways (for example, incineration).

- Any liquid waste.
- Infectious medical or veterinary waste.
- Whole or shredded used tyres.
- Waste that might cause a problem in the landfill (such as hot or chemically active waste).
- Any waste that does not meet the waste acceptance criteria for that class of landfill.

The term **difficult waste** has come into common use and applies to wastes that require handling in a particular way. Examples of difficult waste are shown below.

- **Invasive plants** that are waste materials, both soil and plant matter, contaminated with Japanese knotweed or giant hogweed, which can only be disposed of at sites that are specifically licensed to receive them.
- **Contaminated soil** that is a mixture of soils, stones, rubble and polluting substances, which could be hazardous depending on thresholds, and could be a range of things left over from former use of the site.
- **Gypsum and plasterboard wastes** that, when mixed with biodegradable waste, can produce hydrogen sulphide gas in landfill, which is both toxic and odorous.

Gypsum and plasterboard waste

The land filling of gypsum and other high sulphate-bearing wastes, together with biodegradable waste, has been prohibited. From 1 April 2009 all gypsum waste should be segregated from biodegradable waste before being sent to landfill. It should be noted that plasterboard waste itself (unless contaminated) is not hazardous waste but must be segregated.

3.5 Waste hierarchy

Article 4 of the revised WFD (2008/98/EC) requires that all reasonable measures should be taken to prevent waste and consider the waste hierarchy when waste is transferred.

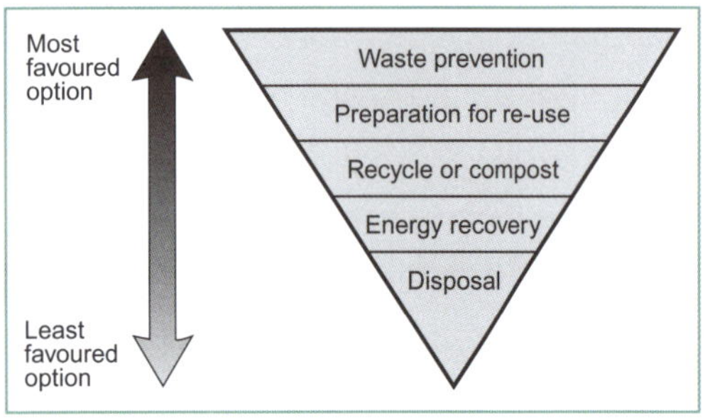

Waste hierarchy

The waste hierarchy is a series of steps for dealing with waste in order of priority and signifies the relative environmental benefits that can be made at each stage.

The waste hierarchy shows that the highest priority is waste reduction or prevention so that the need for other options (such as reuse, recycling and energy recovery) would be dramatically reduced.

The Waste (England and Wales) Regulations now require that you declare on the waste transfer documentation that you have considered the waste hierarchy in the management of the waste.

Examples of applying the waste hierarchy on a construction project for waste reduction could include the following.

- ☑ Prevention or minimisation.
  - Designing the project to suit standard product sizes and to avoid site cutting.
  - Designing the project levels to avoid excavation materials going off site.
  - Specification to allow the use of recycled materials.
  - Pre-assembling components off site or using pre-cast sections.
  - Not over-ordering materials.
  - Reducing the amount of packaging.
  - Ordering materials at the size required, to avoid off-cuts.
  - Promoting employee awareness of environmental matters.
  - Requiring sub-contractors to have a waste management policy.
  - Not over-excavating.
- ☑ Reuse.
  - Reusing soil for landscaping.
  - Using off-cuts of timber for alternative uses.
  - Using brick rubble as hard-core.
  - Investigating local environmental or construction work where other materials might be reused.

Office recycling

Plasterboard recycling

Timber segregation

Segregated waste streams

- ☑ Recycling.
  - Crushing waste concrete to use as hardcore.
  - Recycling asphalt planings as road sub-base or temporary surfacing.
  - Recycling scrap metal, glass and waste oil.
  - Segregating waste materials to a separate well-planned area.
  - Recycling timber off site for educational use or social enterprise.
  - Recycling office waste (such as paper, cans and plastics).
- ☑ Recovery.
  - Sending waste for composting or for energy recovery (such as timber off site to be shredded for use as biomass fuel, or composting).
- ☑ Disposal.
  - Sending canteen or office waste for disposal to a local landfill site.

The WRAP website contains a number of useful resources for designing out waste, managing waste, what materials can be recycled and organisations for doing this.

The recycling or treatment of waste on or off site is likely to require a waste management permit (England and Wales) or licence (Northern Ireland and Scotland), or registered exemption. (*Further details on waste management permitting are outlined in 3.9.*)

Social enterprise for waste – National Community Wood Recycling Project

The NCWRP was founded in 2003 to help set up and develop a nationwide network of wood recycling social enterprises.

Modelled on the multi-award-winning Brighton and Hove wood recycling project, the aim of these social enterprises is to:

- ☑ save resources by rescuing and reusing waste timber that would otherwise be landfilled (or at very best down cycled into woodchip)
- ☑ create sustainable jobs and training and volunteering opportunities for local people – especially those who might be described as disadvantaged and find it difficult to get back to employment.

The NCWRP rescues around 11,500 tonnes of wood every year. More than 40% of this is reused.

The NCWRP website provides further details of the project, partners and locations of where wood is being recycled.

3.5.1 Treatment of waste

The Environmental Permitting Regulations (England and Wales), Landfill Regulations (Northern Ireland) and the Landfill Regulations (Scotland) state that non-hazardous waste must now be treated before being sent to landfill. In practical terms treatment is applying the waste hierarchy to reduce the quantity of waste that ends up in landfill.

Treatment must satisfy all three criteria of a three point test, as shown in the table below.

1.	Be a physical, thermal, chemical, or biological process (including sorting).
2.	Change the characteristics of the waste.
3.	Be carried out in order to: – reduce the volume of the waste – reduce the hazardous nature of the waste – facilitate handling of the waste – enhance recovery of the waste.

Segregated waste streams

On construction sites, in practical terms, this can be achieved by setting up appropriate segregated skips and separating out **(sorting)** any wastes that can be reused or recycled, which will change the **characteristics of the original waste stream**. This in turn will aid in **reducing the volume** of the waste, **facilitating the handling** of the waste and **enhancing recovery** of the waste destined for landfill.

Hazardous wastes are required to be stored and disposed of separately from non-hazardous wastes. Separating these out will **reduce the hazardous nature** of the original waste stream.

Sending your waste to a transfer station or recycling facility, for sorting and recovery prior to the residual waste being sent to landfill, will also satisfy these treatment requirements.

Excavated materials that are to be treated on or off site are generally considered to be waste and the treatment facility operator must have an appropriate environmental permit or register a waste exemption allowing that particular treatment of the excavated materials. *(For further explanation of this issue refer to 3.9 and 3.10.)*

3.5.2 Waste (Scotland) Regulations

These regulations apply to all waste producers, including construction companies, by expanding their duty of care to include a requirement to present dry recyclable material for separate collection. To ensure waste producers can comply with their duty of care, it is necessary for the waste industry to provide services that enable the separate collection of dry recyclables from 1 January 2014. This includes materials such as glass, metals, plastic, paper, card or cardboard.

Food businesses, in urban areas, producing large amounts of food waste (> 50 kg per week) are required to present that food waste for separate collection from 2014. Smaller food businesses have until 2016 to comply with the duty or, if producing very small amounts of food waste (< 5 kg per week), are exempt from the duty altogether. Construction canteens come under the definition of a food business.

The use of macerators to dispose of food waste in the sewer system will also be banned from 1 January 2016, except for domestic premises and food producers in rural areas.

3.5.3 Demolition and refurbishment information data sheets

The demolition and refurbishment information data sheets (DRIDS) have been developed to help identify various waste streams and explain how they can be reused and recycled. They have been developed by the demolition industry, by giving consideration to relevant knowledge and practicable information available on materials they currently manage, or will be required to manage, in the future.

The design and construction of buildings, structures and infrastructures are now governed by tighter regulation and legislation. Therefore, the use of DRIDS allows users to be better informed about the types of materials and products they will encounter and how best to maximise environmental and economic gain.

There are 12 **DRID groups** plus a miscellaneous group, each with their own distinct shape, colour and code, to ease recognition and be inclusive. Each group contains a number of DRIDS, depending on their material or product make-up. For example, the wood group includes DRIDS that are generally one product, such as plywood, chipboard or dimensional timbers. However, wood furniture or wood framed glass panel systems may be in the composites group. The colours have been chosen to reflect those used by the Building Research Establishment (BRE), WRAP, Institution of Civil Engineers (ICE) and Zero Waste Scotland.

Examples of DRID groups

Code	Title	Code	Title
C1	Structural insulated panel	M1	Steel
F1	Vinyl floor tiles	M2	Copper
F2	Carpet tiles	M5	Aluminium
G1	Plasterboard	P1	Plastic pipes
H2	Fluorescent tubes	W1	Dimensional timber
I1	Glass	W2	Chipboard
I2	Concrete	W3	Plywood
I3	Bricks	Z2	Foam insulation

DRIDS are a useful resource for the wider construction industry.

www.

For further information on DRIDS refer to the National Federation of Demolition Contractors or download the latest app.

3.6 Duty of care and waste carrier registration

If you are involved in managing waste you have a legal duty of care. The duty of care applies to everyone involved in handling the waste, from the person who produces it to the person who finally disposes of or recovers it.

Duty of care is one of the main ways to combat fly-tipping and means that:

- ☑ waste has to be stored in a secure location and measures taken to prevent its escape
- ☑ waste has to be passed to an authorised person holding a valid licence or permit
- ☑ any waste passed to an authorised person must be supported by the relevant documentation.

3.6.1 Registration of waste carriers, brokers and dealers (England and Wales)

The Waste Regulations (England and Wales) implements a system for the registration of:

- ☑ waste carriers (those who move waste)
- ☑ waste brokers (those who arrange the movement and/or disposal on behalf of others)
- ☑ waste dealers (those who use an agent to buy and sell waste).

It is possible for a single registration to cover all of this work.

There are two classes of registration known as upper tier and lower tier.

3.6.1.1 Lower tier

Those registered in the lower tier are known as **specified persons**. This group includes all those who are currently registered as professionally exempt. In general this refers to waste authorities, charities, voluntary organisations and those who only manage wastes from agricultural premises, animal by-product wastes or wastes from mines or quarries.

Anyone who normally and regularly carries their own business waste (excluding construction and demolition waste) will need to register in the lower tier.

There is currently no fee for lower tier registration; the registration will be valid until revoked or cancelled.

3.6.1.2 Upper tier

There will be a fee payable for upper tier registration. Upper tier registration will be valid for three years unless revoked or cancelled. If you are a waste carrier, broker or dealer but you are not a specified person you will need to register in the upper tier. All those who are currently registered and not professionally exempt will automatically be transferred to the upper tier. **This includes those who carry their own construction and demolition waste. The existing renewal date will remain valid.**

The same system of registration applies in Northern Ireland under the Controlled Waste (Registration of Carriers and Seizure of Vehicles) Regulations (Northern Ireland).

www.

An application for registration as a carrier, broker or dealer can be made on the Environment Agency, Natural Resources Wales or Northern Ireland Environment Agency websites or by downloading the appropriate form.

3.6.2 Registration of waste carriers (Scotland)

The Controlled Waste (Registration of Carriers and Seizure of Vehicles) Regulations in Scotland require that if you transport waste within the UK, in the course of your business or in any other way for profit, you must register as a carrier of waste with the local waste regulation authority.

You must register even if you only carry your own company's waste or carry waste on an infrequent basis. This applies whether you are a self-employed contractor, part of a partnership or a company.

Registration only needs to be made in the area where your company has its head office. All other offices will be covered by this one registration.

Application for registration as a waste carrier must be made on the prescribed form, which is obtainable from the local waste regulation authority office.

If you have applied for registration but have not yet received the documentation, you will be deemed to be registered and may carry waste.

!

The law demands that waste carriers keep a copy of their registration document on their vehicle. Do not accept photocopies of registration documents as proof of registration. If you have any doubt as to whether the carrier is registered or not, ask the appropriate waste regulation authority.

3.7 Controlled waste and transfer notes

All waste subject to the provisions of the Environmental Protection Act is known as controlled waste and includes waste from domestic, commercial and industrial premises as well as hazardous waste.

Under the duty of care waste must be passed to an authorised person (a holder of a waste carrier or waste management permit or licence).

When non-hazardous waste is transferred to an authorised person this must be supported by a waste transfer note. This also applies even if you have produced and carry waste yourself. The transfer of hazardous waste must be supported by a consignment note *(refer to 3.8)*.

Refer to Appendix B for a sample of a controlled waste transfer note.

Whenever you pass non-hazardous waste on to someone else, you will have to declare on the waste transfer note that you have applied all reasonable measures to apply the waste management hierarchy.

It is a requirement to include the appropriate 2007 SIC code on all waste transfer notes.

The controlled waste transfer note must:

- ☑ include a declaration that you have taken all measures to apply the waste management hierarchy
- ☑ include the appropriate SIC code, identifying the area of construction that generated the waste
- ☑ give a description of the waste, including the six-digit EWC code
- ☑ state the quantity
- ☑ state how the waste is contained, whether loose or in a container and, if in a container, the kind of container
- ☑ identify the carrier of the waste
- ☑ state the date, time and location of transfer
- ☑ identify the disposal site
- ☑ contain your signature and the signature of the authorised person receiving your waste.

A copy of the transfer note must be kept for a minimum of two years. A copy must also be given to the disposal site representative, who will sign the documents to say where, when and how the waste will or has been disposed of.

If you use a registered carrier to dispose of your waste for you, the transfer note must contain all the points described above and, in addition, state the following.

- ☑ Name and address of the carrier, their licence registration number and issuing authority.
- ☑ Place of transfer.

If you use a registered carrier to remove your waste, they will raise and distribute the necessary documentation and will:

- ☑ give you a copy to keep
- ☑ keep a copy for themselves
- ☑ deliver your waste and a copy of the document to the management at the disposal site.

You must keep all controlled, non-hazardous waste transfer documentation for two years.

The electronic duty of care (EDOC) is a new non-mandatory national system that can allow users to create, share, sign and register waste transfers online in electronic format, rather than using paper-based waste transfer notes. It cannot currently be used for the transfer or consignment of hazardous waste.

For further details of the electronic duty of care system refer to the EDOC website.

3.8 Hazardous waste and consignment notes

From 1 April 2016, premises that produce, plan to produce or store 500 kg or more of hazardous waste in England no longer need to register as hazardous waste producers. (The requirement still exists for sites in Wales.) This change affects the entire hazardous waste control system.

Sites producing or storing less than 500 kg per year must still comply with the Hazardous Waste Regulations (England and Wales). There is no requirement to register your site in England, Northern Ireland or Scotland.

For further information refer to the Government legislation website.

E 03

The Hazardous Waste Regulations in England, Northern Ireland and Wales, and the Special Waste Regulations in Scotland, require that all hazardous waste must be segregated from non-hazardous waste. The following controls should be adopted to comply with the regulations.

Segregated hazardous waste

- ☑ Different types of hazardous wastes should be segregated, as you will clearly need to identify the quantities and types on the hazardous waste consignment note.
- ☑ Mixing of different types of hazardous waste should be avoided as this may inadvertently create an explosive or fire risk, particularly in warm weather.
- ☑ The mixing of hazardous waste with non-hazardous waste to dilute the material below the threshold concentration is banned.
- ☑ Packaging or containers contaminated with hazardous substances should be treated as hazardous waste unless it can be shown that the concentration (including the packaging) is below the threshold limits.

Where individual products are combined to form a substance (such as adhesives and resins) then each component should be considered for its hazardous properties and disposed of accordingly. Resins are often inert when set so leaving materials to dry before disposal may make them non-hazardous.

3.8.1 Removal of hazardous waste

If you wish to have hazardous waste removed from site, a document called a **hazardous waste consignment note** (special waste consignment note in Scotland) must be prepared.

The consignment note must be prepared by the person who is originating the transfer of the waste. In Scotland, a special waste consignment note must be obtained from the local office of the Scottish Environment Protection Agency (SEPA).

The hazardous waste consignment note must include the following information.

- ☑ The premises code (where the site is required to register) (Wales only).
- ☑ A consignment note code*.
- ☑ Address of the producer.
- ☑ Name and address of the consignee (where the waste will be taken).
- ☑ Details of the process that produced the waste.
- ☑ The appropriate 2007 SIC code.
- ☑ A description of the waste, including the six-digit EWC code.
- ☑ The quantity
- ☑ Various details on the properties of the waste, including concentrations, hazard code and proper shipping name.
- ☑ The type of container.
- ☑ The date and time of transfer.
- ☑ Your signature (consignor) and the signature of the authorised person carrying the waste.
- ☑ A declaration that you have taken all measures to apply the waste management hierarchy.

**The consignment note coding system for England and Wales was changed on 1 April 2016. The consignment note code is a unique code for each consignment of hazardous waste and is always in the same format of six letters or numbers, followed by a forward slash, and then five further letters or numbers (for example CJTILE/A0001).*

3.8.2 Coding format for hazardous waste consignment notes

A consignment note code must be created (in the format XXXXXX/YYYYY) and entered onto the consignment note. The table below shows the coding format for hazardous waste consignment notes in England and Wales.

<table>
<tr><th colspan="3">Coding format for hazardous waste consignment notes</th></tr>
<tr><th>Code</th><th>Explanation</th><th>Example</th></tr>
<tr><td>XXXXXX/</td><td>First six letters and/or numbers (not symbols or spaces) of the name of the company entered in Part A2 of the consignment note. This must be followed by a forward slash '/'.
Note: if your company name has less than six letters and/or numbers you must assign the letter 'Q' to the remaining characters.</td><td rowspan="2">CJTILE/A0001</td></tr>
<tr><td>YYYYY</td><td>Exactly five digits made up of numbers and/or letters (not symbols or spaces) of your choice.
Note: each code can only be used once from the address in Part A2. You must change the 'YYYYY' to create a different code each time waste is consigned from that premises.</td></tr>
<tr><th colspan="3">Additional letters</th></tr>
<tr><td colspan="3">An additional letter must be assigned at the end of the consignment note code for certain types of consignment:
'V' to waste removed from ships 'XXXXXX/YYYYYV'
'F' to fly-tipped waste 'XXXXXX/YYYYYF'
'D' to waste moving under a consignee return derogation 'XXXXXX/YYYYYD'
'P' to continuous piped waste.
Additional letters must also be assigned to rejected loads.
The consignment note form contains two extra grey boxes for these additional letters.</td></tr>
</table>

In Northern Ireland and Scotland, a copy of completed consignment notes must be received by NIEA or SEPA at least 72 hours before the waste is due to leave site. Completed consignment notes are valid for 28 days after the anticipated date of collection.

You must keep all hazardous waste documentation for three years.

For an example of a hazardous waste consignment note refer to Appendix C.

3.9 Environment permits and exemptions

3.9.1 Environmental permitting in England and Wales

The Environmental Permitting Regulations (England and Wales) (EPR) combine the system of waste management licensing previously regulated under the Waste Management Licensing Regulations, and the system of permitting installations in the Pollution Prevention and Control Regulations (England and Wales). The EPR also now include the provision of permits to deal with groundwater protection, water discharges and flood defence activities.

For further information on environmental permits refer to the Government website.

For further information on water protection refer to Chapter E05 Water management.

The EPR specify which waste work requires an environmental permit and allows some waste operations to be exempt from requiring a permit (as in Schedule 3 discussed below). Certain waste operations covered by other legislation are excluded from these permitting arrangements.

An environmental permit is required for seven different classes of **regulated facility**, five of which relate to waste (the other two cover groundwater and water discharges).

- ☑ **Installations.** Generally these are facilities at which industrial, waste and intensive farming falling (mainly) under the Integrated Pollution Prevention and Control Directive are carried out. These include landfill sites.
- ☑ **Waste operation.** Any other waste activity that is not defined as an installation will be classed as a waste operation. This includes the depositing, treatment or recycling of waste that is not exempt under the Environmental Permitting Regulations (such as waste transfer stations) including the treatment of contaminated land.

19/07/18

Environmental Protection Agency (SEPA)

Natural resources Wales (NRW)

Holiday Inn

Northern Ireland Environment agency (NIEA)

environment agency (EA)

1

Raise awareness of enforcement → local authorities

construction polluting environment → earth, water, air.

energy consumed by glass, brick, iron...

(construction standards (eg. BREEAM, Code of Sustainable homes, etc)

Betari Cycle

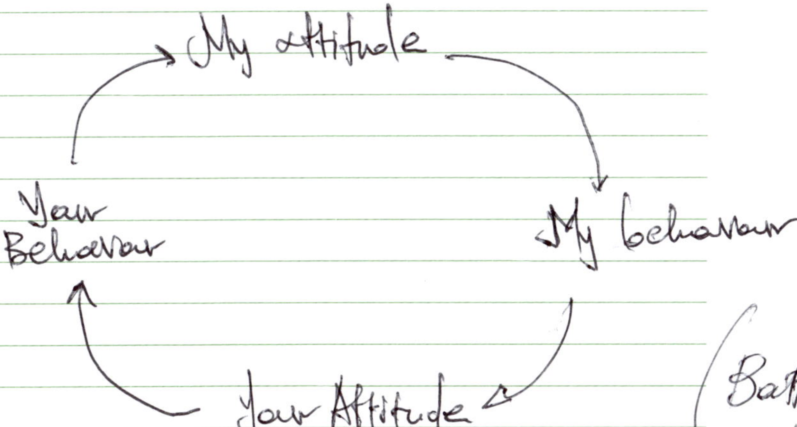

(Battle of Dan
ock 100 years)

~~Sewer~~ Sewers & Treatment

shower
toilet

Wash vegetables
&

Mobile plant. Mobile ...

...eration (such as crushers).

...aste facility.

... apparatus) or the

...s.

...nit. You ...is offence.

...obile treatment plant ... have to be met to comply

...ration.

... to be registered

...es it, and then to

... Part A(1) work that ... Authorities. The EA ... d air), whereas Local

...ed with construction.

...permit

... a standard permit ... specific deployment

L2

Who polices the environment?

Secretaries of State (DCLG, DBEIS, DEFRA)

concrete baths for safety of diesel tanks → 10% bigger

Environmental Management System

14001 Environmental Standards

9000/9001 Quality

Risk Assessment

- Identify hazards & risks
- Implement controls
- Typical Controls spill kits etc like PPE but for environment
- Ask questions

Environmental Aspect → Impact

(cause) → (effect)

Pollution pathways

(source) → (pathway) → (receptor)

pollution origins — way pollution reaches

oil energy — ground factory — fish dead global/whole atmosphere

Holiday Inn
holidayinn.co.uk or 0800 40 50 60

... soil and/or contaminated waters requires a mobile treatment permit (MTP).

A mobile treatment permit is used to regulate a mobile plant activity that involves treatment either in situ or off site. The permit sets out the type and extent of work that can be carried out. A site-based permit has to be used where a mobile plant permit is not applicable. The environmental permit can be either a standard rules permit or a bespoke one depending upon the type of treatment and site location.

Operators who want to treat contaminated soil and/or contaminated waters using their mobile plant permit at a particular site must submit a site-specific deployment application (the deployment form and supporting information). The deployment application details site-specific information and potential impacts arising from the proposed use of the mobile plant. The operator must demonstrate that the activity will not cause pollution of the environment, harm to human health or serious detriment to local amenities.

Following treatment under a MTP the material would normally cease to be waste, providing it was excavated and treated on the site where it was used or is part of a remediation cluster. Where this is not the case then an environmental permit or registered exemption would be required.

For flowcharts providing further guidance on permitting the use or reuse of soils or aggregates refer to Appendix D.

The regulator has four months from receipt of the permit application and all supporting information to make a determination. You have the right of appeal to the Secretary of State should your permit not be granted, or the permit is granted but you are not happy with the conditions that have been imposed. Pre-application discussions with the regulator can help in improving the quality of permit applications so early contact with them is advisable.

The Waste (England and Wales) Regulations implement a new permit condition to require waste to be managed in accordance with the waste hierarchy.

Only a person who is in control of a regulated facility may obtain or hold an environmental permit. This person is called the operator.

To obtain an environmental permit you may have to prove that you have the appropriate technical competence (see below) to be able to carry out the relevant work and fulfil the obligations of an operator.

The regulator will consider the following.

- ☑ Whether your management systems are adequate.
- ☑ If your site is run by someone who is technically competent (see below).
- ☑ Any convictions that you, or other persons in your business, may have for pollution offences.
- ☑ If you have taken steps to meet the possible costs of the duties of the permit.

3.9.1.2 Technical competence

Certain types of waste management activity require the operator to demonstrate technical competence. They are called relevant waste operations. These include, for example, landfill sites, transfer stations and certain work involving hazardous waste.

You will be able to show that you are a technically competent operator if you can satisfy one of the following.

- ☑ Compliance with an approved industry scheme. There are currently two approved schemes for operators of relevant waste operations, the:
  - – Chartered Institution of Wastes Management (CIWM) or Waste Management Industry Training and Advisory Board (WAMITAB) scheme
  - – Environmentally Sensitive Areas/EU scheme.
- ☑ Holding an appropriate certificate of technical competence (CoTC) from the WAMITAB, which issues a range of certificates for the managers of most types of waste management site. This can be checked on the WAMITAB CoTC database.
- ☑ Holding a registered and validated deemed competence status (see below).
- ☑ You have previously completed an environmental assessment for non-CoTC work.

Persons operating under a waste disposal licence before May 1994 have been deemed to be technically competent for that operation, and have never been required to demonstrate their technical competence. These individuals will retain deemed competence status and will now be assumed to have passed the continuing competence assessment.

3.9.1.3 Exemptions from environmental permitting in England and Wales

There are a number of exemptions from environmental permitting for certain waste activities that are not seen as a threat to the environment. The general requirements and descriptions of these exemptions are set out in Schedule 3 of the Environmental Permitting Regulations.

The Environmental Permitting Regulations have significantly changed the descriptions and references for exemptions, which are now grouped into four categories.

Refer to Appendix E for materials and quantity thresholds for U1 or T5 exemptions.

1. **Use of waste.** This includes the recovery or reuse of waste for a purpose. These exemptions will have a 'U' reference (for example, *U1 – Use of waste in construction*).
2. **Treatment of waste.** This includes the treatment of material for the purpose of recovery. These exemptions will have a 'T' reference (for example, *T5 – Screening and blending of waste for the purposes of producing an aggregate or soil and associated prior treatment*). Note that some wastes produced under this exemption may then be used under another exemption (such as the production of aggregates) that is then used under a *U1 – Use of waste in construction exemption.*
3. **Disposal of waste.** This includes the disposal of certain types of waste onto land. These exemptions will have a 'D' reference (for example, *D1 – Deposit of waste from dredging of inland waters*).
4. **Storage of waste.** This includes the storage of waste at a location other than where it was produced pending its recovery or disposal. These exemptions will have an 'S' reference (for example, *S2 – Storage of waste in a secure place*).

These exemptions generally exclude hazardous waste and there will be conditions on most exemptions granted. Some examples of the conditions included in the exemption are shown below.

- ☑ Limits on the quantities and time periods for the temporary storage of waste produced on site for reuse on that site.
- ☑ Limits on the quantities for the spreading, on land, of waste soil.
- ☑ That certain other precautions must be taken.

You may deal with waste under an exemption subject to the conditions imposed, but you must make sure that you do not pollute the environment or cause harm to anyone's health.

To obtain a waste management exemption under Schedule 3 of the Environmental Permitting Regulations, an application must be made, together with the appropriate fee, to the correct waste regulation authority. In most cases this is the EA or NRW.

However, if you want to crush bricks, tiles and concrete you may need to register a *T7 – Treatment of waste bricks, tiles and concrete by crushing, grinding or reducing in size* exemption with your Local Authority. Also note that the equipment will need to have a valid mobile plant permit, which must be obtained from the Local Authority in which the business is situated.

The information required varies considerably depending on the relevant exemption applied for. Notifications for more complex exemptions will generally require a form to be completed and must be supported by the following type of information.

- ☑ Name and address of applicant for the exemption.
- ☑ Location of the site where the waste activity is being carried out.
- ☑ Details of relevant planning permissions.
- ☑ Description, type and analysis of the waste.
- ☑ Intended use for the waste.
- ☑ The relevant fee.

A number of the exemptions also have time constraints and cannot be renewed before the expiry date. For example, under a *U1 – Use of waste in construction* exemption, you cannot register the exemption more than once at any one place during the three-year period from first registration.

Paragraph 103 of the Environmental Permitting Regulations deals with transitional arrangements for existing waste exemptions that were in place before the regulations came into force on 6 April 2010.

For flowcharts providing further guidance on exemptions from permitting for the use or reuse of soils or aggregates refer to Appendix D.

3.9.2 Waste management licensing in Northern Ireland and Scotland

In Northern Ireland the waste licensing system is created under Part II of the Waste and Contaminated Land Order (Northern Ireland) and controlled through the Waste Management Licensing Regulations (Northern Ireland).

In Scotland, the existing waste management licensing system is controlled under Section 33 of Part II of the Environmental Protection Act and the Waste Management Licensing Regulations (Scotland).

A waste management licence is required if you deposit, recover, treat or dispose of controlled waste. If you do any of these things without a licence, or a licence exemption, you could be fined and/or sent to prison.

Obtaining a full waste management licence can be a lengthy and costly exercise. To obtain a waste management licence you have to prove that you are a fit and proper person. Applications for waste management licences should be made to NIEA or SEPA, who will consider the following.

- ☑ Any convictions that you, or other persons in your business, may have for pollution offences.
- ☑ If your site is run by someone who is technically competent.
- ☑ If you have taken steps to meet the possible costs of the duties of the licence.

3.9.2.1 Technical competence

The Waste Management Licensing Regulations (Northern Ireland) remain in force. There is no change in the technical competence requirements in the province, as a result of the regulatory changes that took place in England and Wales on 6 April 2008, with the enactment of the Environmental Permitting Regulations.

In Northern Ireland the certificate of technical competence (CoTC) remains the primary means for managers of appropriate waste management facilities to demonstrate their technical competence, in accordance with the Waste Management Licensing Regulations (Northern Ireland).

Under the Waste Management Licensing Regulations (Scotland) there is no longer a legal requirement for a technically competent person in Scotland to hold a CoTC. These were previously regulated by the WAMITAB under Waste Management Licensing Regulations.

CoTCs remain an appropriate qualification to demonstrate competence in Scotland and can be used on a voluntary basis. Operators in Scotland should contact their SEPA regulatory officer in the first instance to check other competence arrangements.

For further details of competence requirements refer to the WAMITAB website.

3.9.2.2 Waste management licence applications

An application for a waste management licence can only be made if planning permission has been granted. To obtain a waste management licence, you should apply to the waste licensing office of NIEA or SEPA and ask the following questions.

Questions for NIEA or SEPA

- How do you apply for the licence?
- What information do they require?
- How can you show that you are a fit and proper person, including technical competence?
- How much is the application fee for the type of site you wish to run?

NIEA and SEPA have four months from when all information was received to consider the application. You have the right of appeal to the Secretary of State should your licence not be granted, or the licence is granted but you are not happy with the conditions that have been imposed.

3.9.2.3 Exemptions from waste management licensing in Northern Ireland and Scotland

There are a number of waste management exemptions set out in Schedule 2 of the Waste Management Licensing Regulations (Northern Ireland) and Schedule 1 of the Waste Management Licensing (Scotland) Regulations. These exemptions generally exclude hazardous waste and there will be conditions (examples below) on most exemptions granted.

- Limits on the quantities and time periods for the temporary storage of waste produced on site for reuse on that site.
- Limits on the quantities for the spreading, on land, of waste soil.
- That certain other precautions must be taken.

You may deal with waste under an exemption subject to the conditions imposed, but you must make sure that you do not pollute the environment or cause harm to anyone's health.

To obtain a waste management exemption, an application must be made, together with the appropriate fee, if appropriate, to NIEA or SEPA.

The information required varies considerably depending on the relevant exemption being applied for. NIEA and SEPA make a distinction between simple and complex exemptions and the notification process is different for each type. Notifications for complex exemptions will generally require a form to be completed and must be supported by the following type of information.

- Name and address of applicant for the exemption.
- Location of the site where the waste activity is being carried out.
- Details of relevant planning permissions.
- Description, type and analysis of the waste.
- Intended use for the waste.
- The relevant fee.

A number of the exemptions are also required to be renewed on an annual basis. Further advice should be sought from NIEA or SEPA.

3.10 CL:AIRE Development Industry Code of Practice

CL:AIRE (Contaminated Land: Applications in Real Environments) has developed a Code of Practice (CoP) for dealing with the waste management aspects of contaminated land. This CoP, developed with contributions from the development and remediation industries and the EA, is designed to help developers and construction companies to identify if they are dealing with waste and when waste is fully recovered.

Version 2 of the CoP was launched in March 2011 with an extended scope. The scenarios now covered are shown below.

- Reuse of excavated materials on the site of production (contaminated and uncontaminated).
- Direct transfer of clean, naturally occurring soils between sites.
- Reuse of naturally elevated substances in soils (such as arsenic and lead).
- Cluster projects (multiple reuse at different development sites within a similar timeframe).
- Brownfield to brownfield transfers.
- Fixed soil treatment facilities allowing the release of treated materials to the market place.

The main purpose of the CoP is to achieve good practice across the development industry to:

- ☑ assess whether materials are waste or not
- ☑ determine when treated waste ceases to become waste
- ☑ provide an auditable trail to demonstrate that the Code of Practice has been complied with on each site.

The CoP specifies the implementation of a **materials management plan** (MMP), together with a declaration from a competent qualified person, before the commencement of the works. When the declaration is provided to the EA or NRW, demonstrating that the materials are to be dealt with in accordance with the MMP, the EA or NRW may take the view that the materials, where they are used on site, may cease to be waste. The EA have issued a position statement in regard to the CoP supporting this view but it will be up to the site to demonstrate proper controls are in place.

By complying with this CoP it may be possible to avoid the need to apply for a waste permitting exemption for the use of construction waste (U1) as highlighted above. The CoP also allows the direct transfer of uncontaminated natural excavation materials between projects without the need for a permit.

Further advice should be sought from CL:AIRE with reference to the CoP and for further information on the register of materials and services.

Flowcharts providing further guidance on how the CL:AIRE CoP links with environmental permitting in regard to the use or reuse of soils, can be found in Appendix D.

CL:AIRE Code of Practice

Galliford Try – Ingsbeck flood alleviation scheme

Background

The voluntary CoP developed by CL:AIRE in conjunction with the EA helps determine whether materials are classed as waste. The CoP has recently been updated (CL:AIRE CoPv2) to allow the direct transfer of naturally occurring soil materials.

Ingsbeck flood alleviation scheme is a £11 million development in Wakefield. It is spread across a number of areas and comprises the construction of new flood defence walls, channels, embankments and flood storage areas. Part of the construction involved building a new clay flood defence embankment around residential houses, which was valued at approximately £180,000. This required approximately 4,000 tonnes of clay. By applying the CoP, a materials management plan (MMP) was produced to enable the reuse of this material, which had a number of benefits, shown below.

Reduced operational costs

The use of a large volume of waste material would require the use of a standard rules environmental permit, which takes approximately four months in application and incurs costs of around £6,000 for application, subsistence and surrender, as well as the use of a technically competent manager. However, the production of the MMP took three weeks from production to sign off and only cost £500. This benefited the project by reducing programme time and cost, which significantly decreased the overall project cost by £60,000, approximately 33% of the project value.

Reduced landfill costs

The surplus material would have been destined for landfill as there was no further use on the donor site. By utilising the MMP, 4,000 tonnes of material was diverted from landfill (avoiding a £10,000 landfill gate fee for importing inert material) and further benefiting from an 80 tonne embodied carbon saving.

Reduced use of natural resources

By utilising a recycled material, this avoided having to excavate a finite material from a quarry.

Reduced regulatory effort

The use of the MMP does not require any direct involvement from the EA and NRW regulator. This frees up its resources for deployment on other tasks and allows self-regulation for the industry, whilst minimising impact and protecting the environment.

3.11 WRAP quality protocol for the production of aggregates from inert waste

This protocol is published by WRAP and has been produced by the Quarry Products Association (QPA), the Highways Agency (HA) and WRAP as a formalised quality control procedure for the production of aggregates from recovered inert waste. These are referred to in the document as *recovered aggregates*.

The document has two main purposes.

- ☑ To assist in identifying the point at which the inert waste used to produce recovered aggregates has been fully recovered, ceases to be a waste and becomes a product *(further information on the definition of waste and recovery is given in Section 1 of the document).*
- ☑ To give adequate assurance that recovered aggregate products conform to standards common to both recovered and primary aggregates.

The protocol seeks to ensure that recovered aggregates meet the quality and conformity requirements for European standards for aggregates. If they do then they are likely to be regarded as having been completely recovered and having ceased to be waste at that point. However, whether a substance or object is waste, in any particular situation, must still be determined in the light of all the circumstances, having regard to the aims of the WFD and the need to ensure that its effectiveness is not undermined.

You have a responsibility to check that all relevant documentation is correct to confirm that the protocol requirements have been met.

For further information visit the WRAP and CIRIA websites.

For flowcharts providing further guidance on how the WRAP quality protocol links with environmental permitting for the reuse of aggregates refer to Appendix D.

3.12 Asbestos waste

There are slight differences in the respective legislation for England and Wales, Northern Ireland and Scotland. Contractors who work in any of these areas must comply with the relevant national legislation.

In essence, there is a duty of care put to everyone in the waste disposal chain. Clients, whether they are commercial or domestic, have a responsibility to ensure that the waste is disposed of correctly. You can be prosecuted if waste is fly-tipped. Where asbestos has been removed as part of a project, a prudent contractor would ensure that proof of disposal formed part of the information that was given to the client. In England, where a site waste management plan or demolition audit is utilised it should include details of the disposal of the asbestos waste and the waste carrier.

The principal contractor should ensure that disposal records are passed to the principal designer to ensure that this information is placed in the health and safety file to confirm that the asbestos has been transported and disposed of legally.

3.12.1 Classification

Asbestos waste is almost always hazardous waste (special waste in Scotland). There are two main scenarios to consider.

- ☑ A single type of waste that has been separated.
- ☑ A mixed waste.

For a single type of waste that has been separated, the hazardous waste concentration limit for asbestos in a waste is 0.1%, or more asbestos (weight for weight). All asbestos-containing materials (ACMs) used in building materials will contain asbestos at or above this concentration, and will be classed as hazardous waste.

For mixed waste, where more than one type of waste is present, mixing is prohibited. Dilution caused by mixing cannot be used to justify a non-hazardous classification. The waste containing the asbestos that is mixed with other waste should be assessed separately and would retain its hazardous status.

Examples of hazardous waste include the following.

- ☑ Another waste such as soil, crushed secondary aggregate, or mixed construction or demolition waste containing one or more pieces of ACM, of a size visible to the naked eye.
- ☑ The mixing of a soil from a contaminated area of a site, containing 0.1% or more asbestos fibres, with other soil from the same site in a stockpile; on a gravimetric (weight for weight) basis.

For further guidance on the classification and assessment of waste visit the Government website.

For further information on asbestos refer to the asbestos chapter in Section B.

3.12.2 Management on the site of production

Where asbestos waste has been identified for disposal, details of the type, estimated quantities and disposal options should be included in the project site waste management plan, if a site waste management plan is being used.

A contractor can have a skip on site for temporary storage during work.

Asbestos cement sheets or other asbestos cement products should be bagged or wrapped if they are being taken directly from the place of production to a landfill (for example, not being taken to and stored temporarily at a contractor's premises or a transfer station). Alternatively, waste can be placed in a suitable skip that has been lined with polythene. This must then be wrapped over.

Where a skip is used on site it must not be used for consolidating asbestos waste from other sites, as this would introduce more complex hazardous waste requirements. The site would then be recognised as a hazardous waste transfer station requiring an environmental permit.

Skips containing asbestos should be secured if left overnight. It is considered good practice to bag or wrap waste from internal work to ensure that no asbestos is dropped when the waste is carried outside. If a van is being used to carry small volumes of asbestos waste, the waste must be double bagged and the bags secured (for example, by putting them in a bin with a lid). The waste must also be physically separated from the driver, any passengers or any other equipment or materials that are being transported in the vehicle.

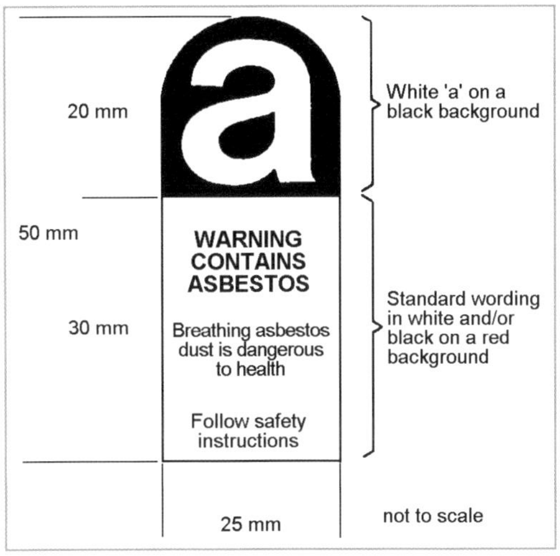

Labelling requirement of asbestos waste

Any fibrous asbestos removed from an enclosure should be double wrapped or bagged. Bagged waste should consist of an inner red bag and an outer clear bag. These should be correctly marked.

The packaging (bagging) and labelling must satisfy the requirements of current legislation and, where asbestos waste is to be transported by road, there is further legislation controlling its safe movement.

Containers should be designed and constructed so as to retain the asbestos without any spillage or loss during normal handling.

Where the asbestos removal contractor uses labels on waste or removed asbestos, the label must be clearly and indelibly printed so that the words in the lower half can be easily read. These words must be printed as specified in the diagram.

The label has to be firmly affixed to the packaging using an adhesive type label, or directly printed onto the packaging (plastic bag).

3.12.3 Removal from the site of production

Contractors must ensure the following for hazardous waste.

- ☑ It is consigned from the place where the waste was produced, using a hazardous waste consignment note.
- ☑ The waste must be transported by a registered waste carrier.
- ☑ In Wales, the hazardous waste producer must also register, with Natural Resources Wales, the place where the hazardous waste was produced (unless they are exempt from registration). This requirement does not apply to England or Scotland.
- ☑ The waste is taken directly from the site of production to a suitably authorised facility within 24 hours. The facility should hold either a relevant environmental permit or an exemption from a permit.

3.12.4 Suitable authorised facilities

Any place where asbestos waste is taken must hold an environmental permit that authorises the acceptance of that waste. This will normally be one of the following.

- ☑ A landfill with a separate cell set aside for asbestos.
- ☑ A temporary storage facility (operated by the asbestos contractor, or a waste contractor) that stores the waste before transferring it to a landfill site.

An asbestos contractor that takes asbestos waste back to their own premises must also hold an environmental permit. Although there is an exemption for storage at a temporary collection point controlled by the producer, this is restricted to bonded asbestos only, and excludes businesses providing a waste management service to customers (such as those contracted to remove asbestos).

- ☑ Asbestos waste must not be put into bays or transferred between different skips or containers.
- ☑ Keep all handling to a minimum (either manual or mechanical). Loading shovels must not be used to move asbestos waste.
- ☑ Keep designated skips secure and locked when asbestos waste is not being deposited into them.
- ☑ Operators must have emergency procedures to deal with delivered, non-conforming asbestos waste. This could include waste that is not properly contained in bags or skips (for example, unwrapped asbestos sheets in open skips). Sites should also have the capability to deal with appropriate non-conforming waste and spillages (for example, broken bags). Records should be kept of non-confirming waste.
- ☑ Do not use a skip that has been used to store asbestos for non-asbestos uses, unless they are verified as clean. Otherwise, the skip must be fully decontaminated.

Any place that receives hazardous waste, including exempt facilities in England and Wales, is a consignee and required to send quarterly returns to the Environment Agency or Natural Resources Wales.

3.12.5 Crushed secondary aggregate

Plants that produce crushed secondary aggregate from construction or demolition waste are not authorised to accept and process waste containing asbestos. They should have effective waste acceptance and screening procedures to prevent this from happening. The secondary aggregates they produce should not contain asbestos.

A secondary aggregate containing:

- one or more piece of ACM (as detailed in Classification of waste), or
- any concentration of asbestos resulting from the processing of pieces of ACM or a waste containing asbestos fibres at a hazardous concentration

is likely to remain hazardous waste and will be subject to waste controls, which would prevent its use in construction.

3.13 Waste electrical and electronic equipment

The Waste Electrical and Electronic Equipment Regulations (WEEE Regulations) first came into effect on 2 January 2007 and apply to England, Northern Ireland, Scotland and Wales. New regulations were introduced in 2014, implementing the recast WEEE Directive 2012/19/EU. These regulations currently apply to ten categories of electrical equipment listed in Schedule 1 and generally cover all types of electrical equipment (for example, computers, power tools and microwaves). Schedule 2 lists the products that fall under each of the ten categories in Schedule 1. However, from 2019 the regulations will cover a wider range of equipment.

A producer selling electrical equipment for non-household use (such as equipment used by a construction company) is obliged to finance the collection, treatment and recycling in an environmentally sound manner of:

- any waste electrical equipment replaced (with equivalent or similar function) by the electrical equipment sold, if it was originally purchased before 13 August 2005, whether supplied by this or another producer
- the electrical equipment the producer sold on or after 13 August 2005 when it is eventually discarded as WEEE.

The collection, treatment and recovery may be undertaken either by the producer, or by their producer compliance scheme (PCS), which they must register with. The PCS should be registered with the appropriate waste regulation authority (EA, NRW, NIEA or SEPA).

Equipment distributors (retailers, wholesalers, mail order or internet dealers) do not have any specific obligations for non-household electrical equipment. However, the PCS registration information should be passed on so that the end user (such as a construction company) can properly dispose of the item at the end of its life.

Over 9,500 tonnes of non-household WEEE goods were collected in the UK in 2016.

3.13.1 End-user responsibilities for WEEE

A construction company's role in the WEEE Regulations means that, to dispose of waste electrical equipment, the following actions must be taken.

Waste electrical equipment

- It must be segregated from other types of waste for disposal.
- If the waste electrical equipment was purchased before 13 August 2005, and is being replaced with new equivalent equipment, then ask the producer for details of its PCS and collection arrangements.
- If the waste equipment is not being replaced with new equivalent equipment, or the PCS cannot be traced, then you must pay to transfer the waste equipment to an approved authorised treatment facility that can accept waste electrical equipment (such as a licensed transfer station).
- If it was purchased after 13 August 2005, then contact the supplier for details of the PCS and collection arrangements (if these have not been provided).
- Any waste transferred to an authorised collector or waste carrier must meet all of the normal requirements for duty of care (such as waste carrier's licence, transfer notes and licensed treatment facilities, for example waste transfer station) approved by the waste regulation authority.

3.14 Waste batteries

The construction industry is a large user of batteries in many types of vehicles, plant and equipment. Currently the recycling rate for portable batteries is low, at around 3%.

The Waste Batteries and Accumulators Regulations came into force on 5 May 2009 and apply to the UK and Northern Ireland. They implement the waste battery provisions of the EU Directive on Batteries and Accumulators 2006/66/EC and set out requirements for waste battery collection, treatment, recycling and disposal for all types of battery.

Waste portable batteries

The regulations deal with three types of battery.

- ☑ **Automotive batteries** used for starting or the ignition of a vehicle engine, or for powering the lights of a vehicle.
- ☑ **Industrial batteries** used for industrial or professional purposes (such as the battery used as a source of power and propulsion to drive the motor in an electric forklift).
- ☑ **Portable batteries** that are sealed, can be hand carried and are neither an automotive battery or accumulator nor an industrial battery. Examples of a portable battery include AA or AAA type battery or the battery used to power a laptop or mobile telephone.

E 03

3.14.1 Take-back of waste batteries

These regulations require that distributors of **portable batteries** (for example, retail stores) have a duty to take back waste portable batteries through facilities such as in-store waste-battery bins. This requirement does not apply to distributors who supply less than 32 kg of batteries per year.

A producer of **industrial batteries** is obliged to provide for the take-back of waste industrial batteries free of charge from the end user in the following circumstances.

- ☑ If the producer has supplied new industrial batteries to that end user.
- ☑ Where, for any reason, the end user is not able to return waste industrial batteries to the supplier who supplied the batteries, providing the waste batteries are the same chemistry as the batteries the producer places on the market.
- ☑ If the end user is not purchasing new batteries, and a battery with the same chemistry as the one being returned has not been placed on the market for a number of years, then the end user's entitlement is to be able to contact any producer to request take-back.

The regulations also require that **automotive battery** producers collect, on request, waste automotive batteries free of charge, from businesses (such as garages, scrap yards, end-of-life vehicle authorised treatment facilities or civic amenity sites, such as Local Authority waste recycling centres) during any calendar year in which the producer places new automotive batteries on the market. Under the regulations producers do not have a duty to collect waste automotive batteries from individual end users.

3.14.2 Guidance for battery users

An important aim of the regulations is to enable end users of industrial batteries to have them treated and recycled at no cost to the end user.

End users of automotive batteries are not entitled to the free collection of their waste batteries by battery producers. Businesses (such as garages, breakdown companies and end-of-life vehicle treatment facilities) are not obliged to accept waste batteries from end users free of charge but may be prepared to do so because these businesses are entitled to a free collection service from battery producers.

As with any waste, the final holders must comply with the duty of care for waste by ensuring that all waste transfers are passed to authorised persons, together with the correct waste transfer documentation. Waste should only be carried by a licensed waste carrier.

www. **For further information on the Waste Batteries and Accumulators Regulations refer to the Government legislation website.**

3.15 Site waste management plans

The requirement for site waste management plans (SWMPs) on construction projects in England became law in 2008 through the Site Waste Management Plan Regulations. The two most important reasons for implementation of the regulations were to:

- ☑ improve resource efficiency and reduce waste
- ☑ prevent fly-tipping.

However, in line with the Government's drive to reduce red tape, **the regulations have been repealed**. From 1 December 2013 SWMPs are no longer mandatory. However, more than two thirds of businesses confirmed that they would continue to use SWMPs even if they are no longer a mandatory requirement. Businesses can still use them as a tool to help identify savings on any project.

For a waste management, storage and disposal checklist refer to Appendix F.

Business benefits of implementing a site waste management plan

A SWMP can provide a number of business benefits.

- Identifying wastes early means they can be minimised through design and procurement practices before construction starts.
- Helping to answer queries from the waste regulation authority simply and easily.
- Helping to avoid prosecution by ensuring that all wastes being disposed of end up in the right place.
- Showing how waste is managed and demonstrating any savings made.
- Managing materials and waste responsibly means there is less risk to the environment.
- Providing valuable information for future projects on the costs and quantities of waste produced (information that can be used to set targets for reduction).

E
03

3.15.1 Preparing site waste management plans

Any client intending to carry out a construction project could prepare a SWMP. It is likely, however, that the client might delegate this responsibility to the principal contractor or contractor.

A successful SWMP requires that the following are, and should continue to be, identified in the plan.

- Client.
- Principal contractor.
- Person who drafted it.
- Nature of the construction work and location.
- Estimated cost of the project.
- Types and quantities of each waste expected to be produced during the project.
- Waste management action proposed for each of these wastes (such as reusing, recycling, recovery or disposal).

The plan should include information on any decisions taken before the plan was prepared relating to the nature of the project, its design, construction methods or materials employed to minimise the amount of waste. The plan should include a declaration from the client and principal contractor that all waste will be dealt with in accordance with the Duty of Care Regulations, and materials will be handled efficiently and waste managed appropriately.

3.15.2 Updating site waste management plans

Where SWMPs are still utilised, it is good practice to maintain certain records.

For projects less than £500,000, where waste is removed, the plan should be updated by the principal contractor or contractor to identify:

- the company or person removing the waste
- the types of waste removed
- where the waste will be taken to.

For projects above £500,000, where waste is removed, the principal contractor should update the plan to identify the:

- company or person removing the waste
- waste carrier's registration number
- waste by providing a description along with a copy or reference to the DoC waste transfer note
- site that the waste is being taken to and whether the operator of the site holds a waste management permit under the Environmental Permitting (England and Wales) Regulations (see above).

As often as appropriate, but at least every six months, the principal contractor should ensure that the plan accurately reflects progress of the project by reviewing the plan and recording the types and quantities of waste produced, reused, recycled, sent for another form of recovery, sent to landfill or disposed of in another manner.

3.15.3 Finalising site waste management plans at project completion

There were also requirements under the Site Waste Management Plan Regulations (now revoked) to add information when the project had been completed. The following information is offered as guidance for consideration to support good practice in relation to SWMPs.

For projects with an estimated value of less than £500,000, within three months of the works being completed the principal contractor had to add the following information to the plan.

- ☑ Confirmation that the plan had been monitored on a regular basis and the works were progressing in accordance with the plan.
- ☑ An explanation of any deviations from the plan.

For projects above £500,000, within three months of the works being completed the principal contractor had to add the following information to the plan.

- ☑ Confirmation that the plan had been monitored on a regular basis and the works were progressing in accordance with the plan.
- ☑ A comparison of the estimated quantities of each waste against the actual quantities.
- ☑ An explanation of any deviations from the plan.
- ☑ An estimate of the cost savings that had been achieved by completing and implementing the plan.

A copy of the SWMP was maintained at the site office or at a location that was easy to access for any contractor who had an involvement within it. Arrangements for the project site waste management plan also needed to be included within site inductions.

Implementing a SWMP can bring a number of business benefits, such as those shown below.

- ☑ Wastes can be identified early and can be minimised through design and procurement practices before construction starts.
- ☑ Queries from the waste regulation authority can be answered simply and easily.
- ☑ It can help avoid prosecution by ensuring that all wastes being disposed of end up in the right place.
- ☑ It shows how waste is managed and can demonstrate any savings made.
- ☑ Materials and waste are managed responsibly and are therefore less risk to the environment.
- ☑ It provides valuable information for future projects on the costs and quantities of waste produced. This information can be used to set targets for reduction.

Whilst the Site Waste Management Plan Regulations did not specify how all of this information should be presented, a standard template and guidance is available from the Government's non-statutory guidance for SWMPs.

www.

For further procedures to help manage reduction of waste refer to Netregs (environmental guidance for businesses in Northern Ireland and Scotland).

E
03

Appendix A – European Waste Catalogue six-digit list of wastes codes

Section 17 – Construction and demolition waste

A six-figure list of wastes code for the type of waste being removed MUST be written on every waste transfer note (for example, skip/muck away tickets).

17 01 Concrete, bricks, tiles and ceramics	
17 01 01	Concrete.
17 01 02	Bricks.
17 01 03	Tiles and ceramics.
17 01 06*	Mixtures of, or separate fractions of concrete, bricks, tiles and ceramics containing hazardous substances.
17 01 07	Mixtures of concrete, bricks, tiles and ceramics other than those mentioned in 17 01 06.
17 02 Wood, glass and plastic	
17 02 01	Wood.
17 02 02	Glass.
17 02 03	Plastic.
17 02 04*	Glass, plastic and wood containing or contaminated with hazardous substances.
17 03 Bituminous mixtures, coal tar and tarred products	
17 03 01*	Bituminous mixtures containing coal tar.
17 03 02	Bituminous mixtures other than those mentioned in 17 03 01.
17 03 03*	Coal tar and tarred products.
17 04 Metals (including their alloys)	
17 04 01	Copper, bronze, brass.
17 04 02	Aluminium.
17 04 03	Lead.
17 04 04	Zinc.
17 04 05	Iron and steel.
17 04 06	Tin.
17 04 07	Mixed metals.
17 04 09*	Metal waste contaminated with hazardous substances.
17 04 10*	Cables containing oil, coal tar and other hazardous substances.
17 04 11	Cables other than those mentioned in 17 04 10.
17 05 Soil (including excavated soil from contaminated sites), stones and dredging spoil	
17 05 03*	Soil and stones containing hazardous substances.
17 05 04	Soil and stones other than those mentioned in 17 05 03.
17 05 05*	Dredging spoil containing hazardous substances.
17 05 06	Dredging spoil other than those mentioned in 17 05 05.
17 05 07*	Track ballast containing hazardous substances.
17 05 08	Track ballast other than those mentioned in 17 05 07.
17 06 Insulation materials and asbestos-containing construction materials	
17 06 01*	Insulation materials containing asbestos.
17 06 03*	Other insulation materials consisting of or containing hazardous substances.
17 06 04	Insulation materials other than those mentioned in 17 06 01 and 17 06 03.
17 06 05*	Construction materials containing asbestos (7).
17 08 Gypsum-based construction material	
17 08 01*	Gypsum-based construction materials contaminated with hazardous substances.
17 08 02	Gypsum-based construction materials other than those mentioned in 17 08 01.
17 09 Other construction and demolition wastes	
17 09 01*	Construction and demolition wastes containing mercury.
17 09 02*	Construction and demolition wastes containing PCB (for example, PCB containing sealants, PCB-containing resin-based floorings, PCB-containing sealed glazing units).
17 09 03*	Other construction and demolition wastes (including mixed wastes) containing hazardous substances.
17 09 04	Mixed construction and demolition wastes other than those mentioned in 17 09 01.

Entries marked with * are either potentially hazardous (hazardous mirror entry) and will require testing to determine if hazardous properties are present or definitely hazardous (absolute) and must be disposed of as hazardous waste.

Appendix B – Example of a controlled waste transfer note

Duty of care: waste transfer note

Keep this page and copy it for future use. Please write as clearly as possible.

Section A – Description of waste

A1 Description of the waste being transferred

List of Waste Regulations code(s)

A2 How is the waste contained?

Loose ☐ Sacks ☐ Skip ☐ Drum ☐

Other ☐

A3 How much waste? For example, number of sacks, weight

Section B – Current holder of the waste – Transferor

By signing in Section D below I confirm that I have fulfilled my duty to apply the waste hierarchy as required by Regulation 12 of the Waste (England and Wales) Regulations 2011 Yes ☐

B1 Full name

Company name and address

Postcode SIC code (2007)

B2 Name of your unitary authority or council

B3 Are you:

The producer of the waste? ☐

The importer of the waste? ☐

The local authority? ☐

The holder of an environmental permit? ☐

Permit number

Issued by

Registered waste exemption? ☐

Details, including registration number

A registered waste carrier, broker or dealer? ☐

Registration number

Details (are you a carrier, broker or dealer?)

Section C – Person collecting the waste – Transferee

C1 Full name

Company name and address

Postcode

C2 Are you:

The local authority? ☐

C3 Are you:

The holder of an environmental permit? ☐

Permit number

Issued by

Registered waste exemption? ☐

Details, including registration number

A registered waste carrier, broker or dealer? ☐

Registration number

Details (are you a carrier, broker or dealer?)

Section D – The transfer

D1 Address of transfer or collection point

Postcode

Date of transfer (DD/MM/YYYY)

D2 Broker or dealer who arranged this transfer (if applicable)

Postcode

Registration number

Time(s)

Transferor's signature

Name

Representing

Transferee's signature

Name

Representing

WMC2A Version 3, August 2011

page 1 of 1

You must keep all controlled, non-hazardous waste transfer documentation for two years.

E
03

Appendix C – Example of a hazardous waste consignment note

Form HWCN01v112

The Hazardous Waste Regulations 2005: Consignment Note

Environment Agency

PRODUCER'S/HOLDER'S/CONSIGNOR'S COPY (Delete as appropriate)

PART A Notification details

1 Consignment note code: /

2 The waste described below is to be removed from (name, address, postcode, telephone, e-mail, facsimile):

3 The waste will be taken to (name, address and postcode):

4 The waste producer was (if different from 2) (name, address, postcode, telephone, e-mail, facsimile):

PART B Description of the waste If continuation sheet used, tick here ☐

1 The process giving rise to the waste(s) was:

2 SIC (2007) for the process giving rise to the waste: . /

3 WASTE DETAILS (where more than one waste type is collected all of the information given below must be completed for each EWC identified)

Description of waste	List of wastes (EWC code)(6 digits)	Quantity (kg)	The chemical/biological components in the waste and their concentrations are:		Physical form (gas, liquid, solid, powder, sludge or mixed)	Hazard code(s)	Container type, number and size
			Component	Concentration (% or mg/kg)			

The information given below is to be completed for each EWC identified

EWC code	UN identification number(s)	Proper shipping name(s)	UN class(es)	Packing group(s)	Special handling requirements

PART C Carrier's certificate

(If more than one carrier is used, please attach schedule for subsequent carriers. If schedule of carriers is attached tick here. ☐**)**

I certify that I today collected the consignment and that the details in A2, A3 and B3 are correct and I have been advised of any specific handling requirements.

Where this note comprises part of a multiple collection the round number and collection number are: /

1 Carrier name:

On behalf of (name, address, postcode, telephone, e-mail, facsimile):

2 Carrier registration no./reason for exemption:

3 Vehicle registration no. (or mode of transport, if not road):

Signature

Date | Time

PART D Consignor's certificate

I certify that the information in A, B and C has been completed and is correct, that the carrier is registered or exempt and was advised of the appropriate precautionary measures. All of the waste is packaged and labelled correctly and the carrier has been advised of any special handling requirements.

I confirm that I have fulfilled my duty to apply the waste hierarchy as required by Regulation 12 of the Waste (England and Wales) Regulations 2011.

1 Consignor name:

On behalf of (name, address, postcode, telephone, e-mail, facsimile):

Signature

Date | Time

PART E Consignee's certificate (where more than one waste type is collected all of the information given below must be completed for each EWC)

Individual EWC code(s) received	Quantity of each EWC code received (kg)	EWC code accepted/rejected	Waste management operation (R or D code)

1 I received this waste at the address given in A3 on: Date | Time

2 Vehicle registration no. (or mode of transport if not road):

3 Where waste is rejected please provide details:

I certify that waste permit/exempt waste operation number:

authorises the management of the waste described in B at the address given in A3.

Where the consignment forms part of a multiple collection, as identified in Part C, I certify that the total number of consignments forming the collection are:

Name:

On behalf of (name, address, postcode, telephone, e-mail, facsimile):

Signature

Date | Time

HWCN01v112

You must keep all hazardous waste documentation for three years.

Appendix D – Waste flowcharts for the reuse of construction materials (soils and aggregates)

The following three flowcharts will help you to decide whether use and reuse of soil and aggregate materials in construction works is a waste activity or not and what you need to do to ensure legal compliance whilst minimising the regulatory burden. These flowcharts may also assist you in completion of a SWMP when assessing reuse or recycling options during design and construction phases. These flowcharts are applicable in England and Wales.

These flowcharts are not intended to cover exhaustive criteria of waste; however they do cover the typical scenarios that are commonly faced on construction and civil engineering projects.

For simplicity, these flowcharts have been split up into three priority operations (*see below and on the following pages*).

EA refers to the Environment Agency in England, and NRW refers to Natural Resources Wales in Wales.

Materials arising from work to be reused on site

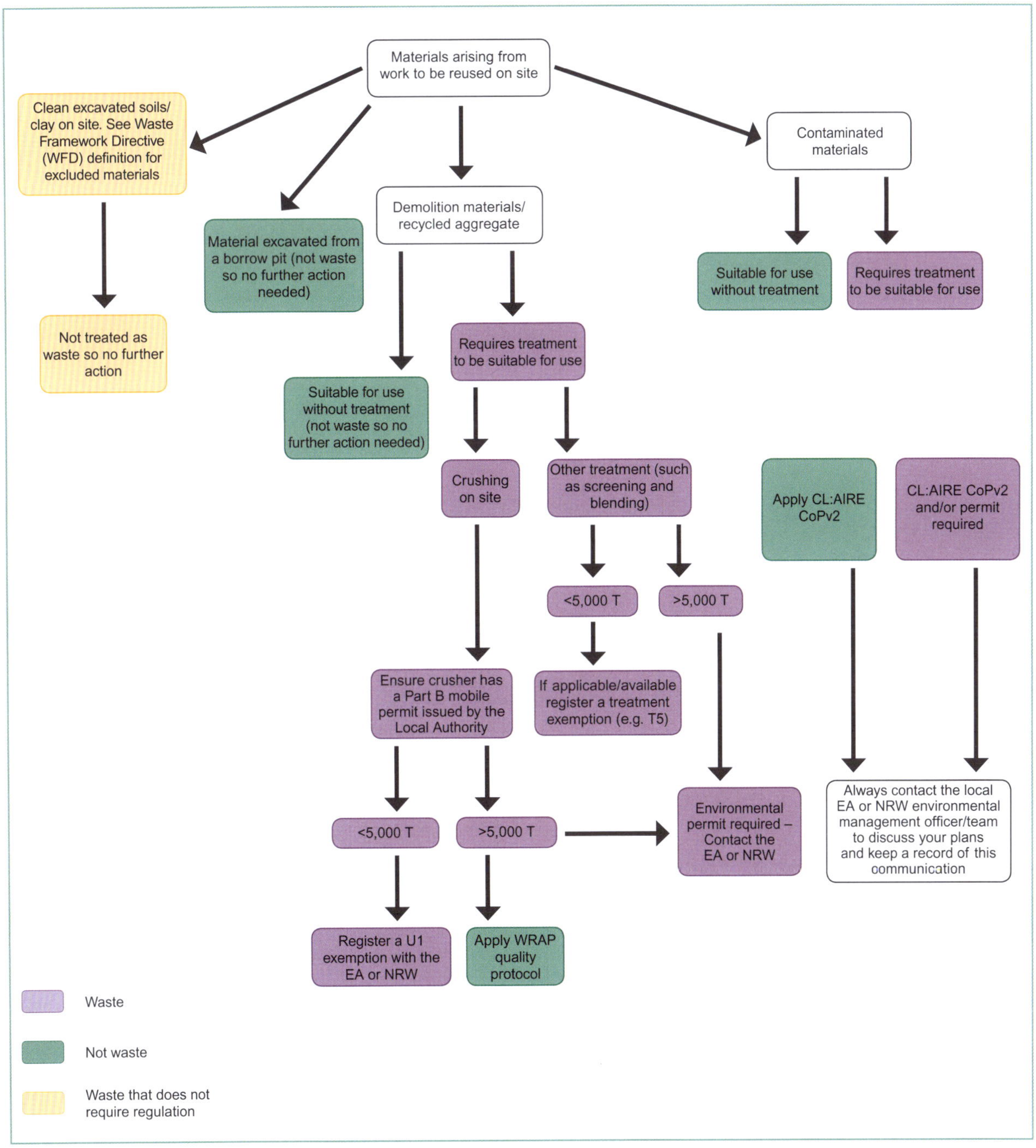

E 03

Materials to be brought onto site

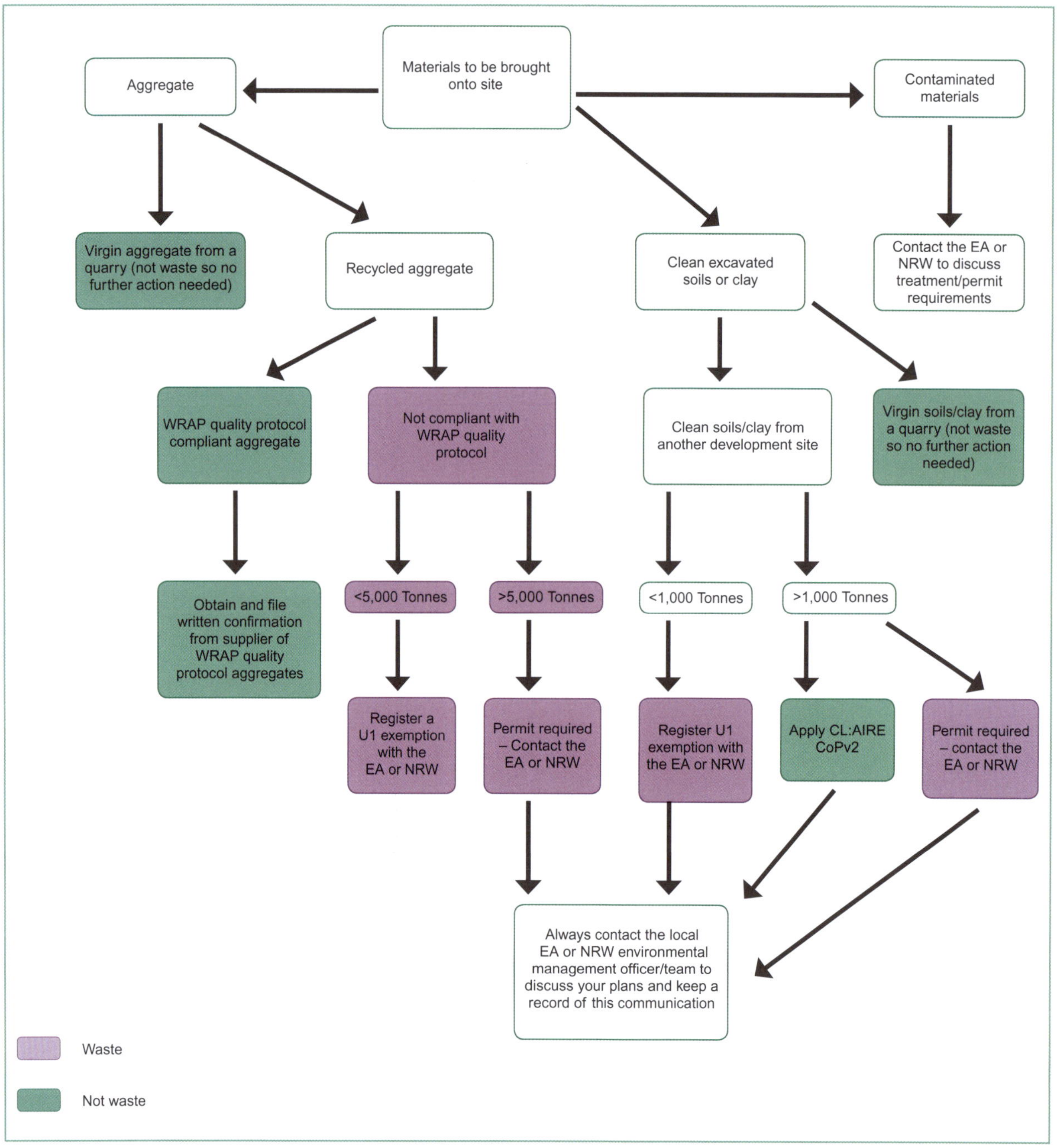

Materials to be sent off site

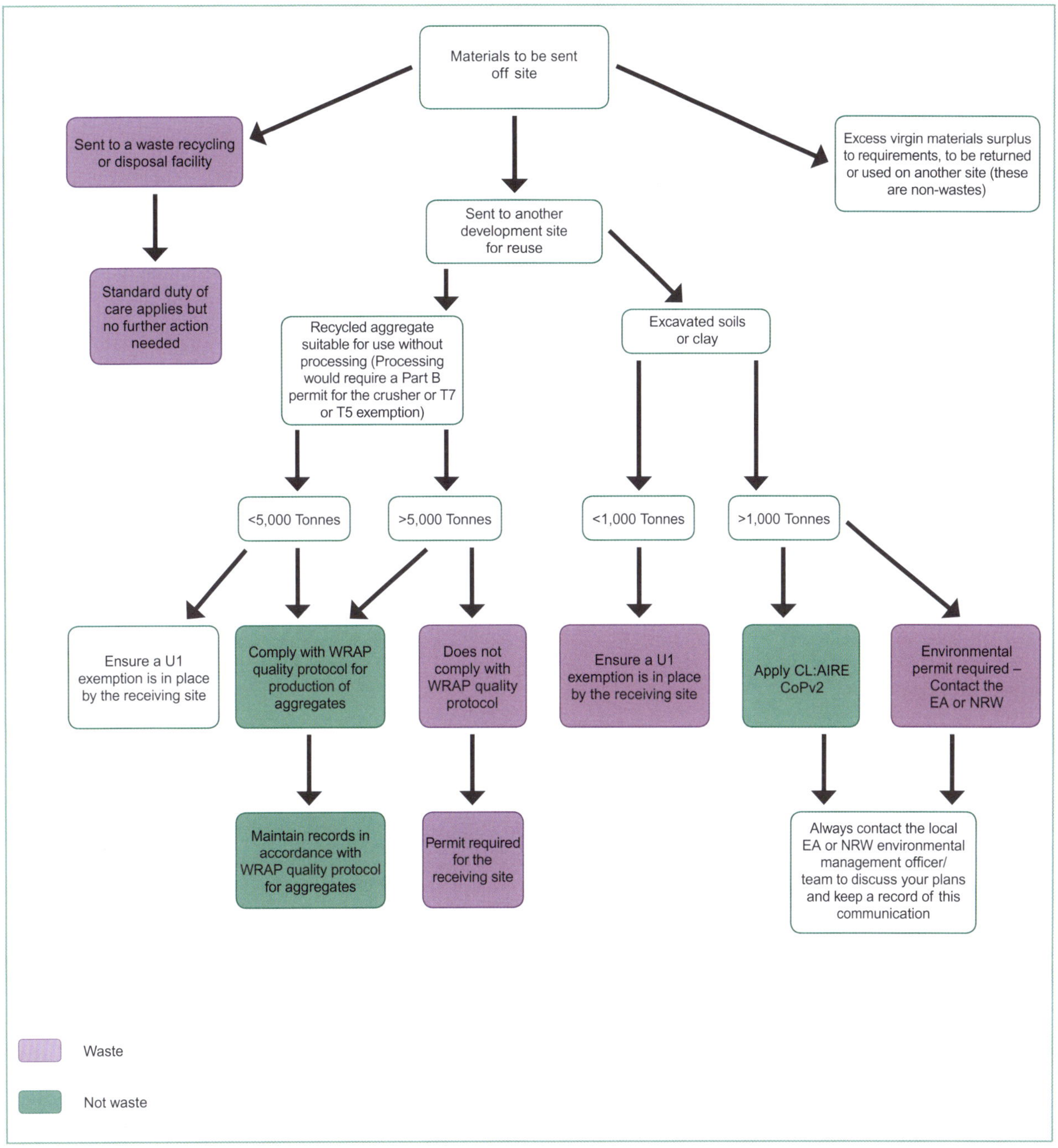

Appendix E – Waste exemption materials and thresholds

Exemption U1 – Use of waste in construction

This exemption allows the use of suitable wastes for small scale construction instead of using virgin raw materials.

Construction for the purposes of this activity means:

> **any building or engineering work, which includes the repair, alteration, maintenance or improvement of an existing work and preparatory or landscaping works.**

Land reclamation is only permissible under this exemption when it's an integral part of a construction activity.

What type of work can you do?

Example work includes:

- ☑ using crushed bricks, concrete, rocks and aggregate to create a noise bund around a new development and then using soil to landscape it to enable grass to grow
- ☑ using road planings and rubble to build a track, road or car park
- ☑ using wood chip to construct a track, path or bridleway
- ☑ bringing in some soil from another place for use in landscaping at a housing development.

Where can you carry out this activity?

This can be done at any place that can comply with the environmental controls listed in the main limits and conditions.

What can't you do?

You can't:

- ☒ treat waste under this exemption to make it suitable for use
- ☒ dispose of waste under this exemption. You can only use waste types that are suitable for use and you should be able to justify the amount of waste that you use
- ☒ use this exemption for land reclamation or disposal in a landfill (you must read the guidance on disposal vs. recovery RGN13: *Defining waste recovery: Permanent deposit of waste on land* and make sure that you are using the waste for recovery only)
- ☒ register this exemption more than once at any one place during the three-year period from first registration
- ☒ de-register this exemption and then re-register it at the same place within a three-year period
- ☒ store the waste for more than 12 months prior to use.

What are the significant limits?

Each table below lists the waste types and quantities that can be used over a three-year period from the date of registering the exemption.

You can use up to:

- ☑ 5,000 tonnes of any single waste stream or any combination of wastes in Table 1
- ☑ 1,000 tonnes of any single waste stream or any combination of wastes in Table 2
- ☑ 50,000 tonnes of any single waste stream or any combination of wastes in Table 3.

What are the significant conditions?

You may use a combination of wastes from Tables 1, 2 and 3 provided you do not exceed the limits for each table. Waste can't be stored for longer than 12 months prior to use.

There are three specific conditions to this exemption relating to certain wastes. These are outlined below and also in the relevant section of *What waste can be used under this exemption?*

- ☑ Any person or company can use up to 1,000 tonnes of dredging spoil for any construction (within the 1,000 tonnes total for wastes from Table 2). Exception is made for the Environment Agency and other statutory authorities carrying out land drainage functions under the Land Drainage Act, the Water Resources Act or the Environment Act. These organisations may use up to 5,000 tonnes of dredging spoil for drainage work (within the 5,000 tonnes total for wastes from Table 1).
- ☑ You can use 1,000 tonnes of wood chip (or similar waste) or road planings to construct tracks, paths, bridleways or car parks only (within the 1,000 tonnes total for wastes from Table 2). The waste must be processed into chipped form prior to use.
- ☑ If you are constructing a road you can use 50,000 tonnes of road planings and road sub-base. The road should be constructed to a specific engineering standard and have a sealed surface in order to qualify for the higher limit.

E
03

What waste can be used under this exemption?

Table 1

You can use up to 5,000 tonnes in total of the wastes below for any construction activity.

Codes	Waste types
01 01 02	Waste from mineral non-metalliferous excavation.
01 04 08	Waste gravel and crushed rock other than those mentioned in 01 04 07*.
01 04 09	Waste sand and clays.
02 02 02	Shellfish shells from which the soft tissue or flesh has been removed only.
10 12 08	Waste ceramics, bricks, tiles and construction products (after thermal processing).
10 13 14	Waste concrete and concrete sludge.
17 01 01	Concrete.
17 01 02	Bricks.
17 01 03	Tiles and ceramics.
17 01 07	Mixtures of concrete, bricks, tiles and ceramics other than those mentioned in 17 01 06*.
17 05 08	Track ballast other than those mentioned in 17 05 07*.
19 12 05	Glass.
19 12 09	Minerals (for example sand and stones).
19 12 12	Aggregates only.

Within the 5,000 tonnes total for use of wastes in Table 1, you can only use the waste below for drainage work carried out for the purposes of the Land Drainage Act 1991(1), the Water Resources Act 1991 or the Environment Act 1995. This is work that can only be carried out by drainage authorities (such as Inland Drainage Boards. Local Authorities or environment agencies).

Codes	Waste types
17 05 06	Dredging spoil other than those mentioned in 17 05 05*.

Table 2

You can use up to 1,000 tonnes in total of the wastes below for construction purposes.

Codes	Waste types
02 03 99, 02 04 01	Soil from cleaning and washing fruit and vegetables only.
17 05 04	Soil and stones other than those mentioned in 17 05 03*.
17 05 06	Dredging spoil other than those mentioned in 17 05 05*.
19 13 02	Solid wastes from soil remediation other than those mentioned in 19 13 01*.
20 02 02	Soil and stones.

Within the 1,000 tonnes total for use of wastes from Table 2, you can only use the waste below for the construction of tracks, paths, bridleways or car parks. The waste must be processed into chipped form prior to use.

Codes	Waste types
17 03 02	Bituminous mixtures other than those mentioned in 17 03 01*.
02 01 03	Plant tissue waste.
03 01 01, 03 03 01	Untreated waste bark, cork and wood only.
03 01 05	Untreated wood including sawdust, shavings and cuttings from untreated wood only.
17 02 01	Untreated wood only.
19 12 07	Untreated wood other than those mentioned in 19 12 06* only.
20 01 38	Untreated wood other than those mentioned in 20 01 37* only.

** For waste codes beginning with 17 refer to Appendix A, for all other codes refer directly to the European Waste Catalogue.*

Table 3

You can use up to 50,000 tonnes in total of the wastes below only for the construction of roads. The roads should be constructed to a specific engineering standard and have a sealed surface in order to qualify for this larger limit.

Codes	Waste types
17 03 02	Bituminous mixtures other than those mentioned in 17 03 01*.
17 05 04	Road sub-base only.

** Refer to Appendix A.*

Exemption T5 – Screening and blending of waste

This exemption allows temporary small-scale treatment of wastes to produce an aggregate or a soil at a place (such as a construction or demolition site).

What type of work does this cover?

Example work includes:

- ☑ screening of soils on a demolition site to remove wood and rubble before sending the soils to a construction site for reuse
- ☑ blending of soils and compost that has been produced under an exemption on a construction site to produce a better soil for landscaping works on that site
- ☑ crushing wastes (except bricks, tiles and concrete) prior to screening or blending
- ☑ grading of waste concrete after crushing to produce a required type of aggregate.

Where can this activity be carried out?

You can treat waste on the site where it is:

- ☑ to be used (for example, on a construction site)
- ☑ produced (for example, on a demolition site).

What can't you do?

You can't:

- ☒ import waste, treat it and then export it elsewhere*
- ☒ treat waste where the main purpose is disposal to landfill or incineration
- ☒ crush waste tiles, bricks or concrete (this comes under a T7 exemption, which must be registered with the Local Authority)
- ☒ treat hazardous waste.

** This applies even if the treated aggregate meets the quality protocol standard and will no longer be considered as waste.*

What are the significant limits?

- ☑ You can store or treat up to 50,000 tonnes of bituminous mixtures for making road stone over a three-year period from the date of registering the exemption.
- ☑ You can store or treat up to 5,000 tonnes of other wastes listed below over a three-year period from the date of registering the exemption.
- ☑ Waste can only be stored for up to 12 months.

What are the significant conditions?

Treatment can only be carried out at the place where the waste is to be used or where the waste is produced. This applies even if the resultant material is no longer considered to be waste.

What else do you need to know?

When you have treated the waste the options available are:

- ☑ if the treated waste meets the requirements of a waste quality protocol (the quality protocols – WRAP) then it will no longer be considered a waste
- ☑ use the treated waste, subject to the significant conditions above, under a use exemption or environmental permit.

What waste can be treated under this exemption?

Codes	Waste types
01 04 08	Waste gravel and crushed rocks other than those mentioned in 01 04 07.*
01 04 09	Waste sand and clays.
02 02 02	Shellfish shells from which the soft tissue or flesh have been removed only.
03 01 01	Untreated waste bark and cork only.
03 03 01	Untreated waste bark and wood.
10 01 01	Bottom ash, slag and boiler dust (excluding boiler dust mentioned in 10 01 04*).
10 01 15	Bottom ash, slag and boiler dust from co-incineration other than those mentioned in 10 01 14.*
17 01 01	Concrete.
17 01 02	Bricks.
17 01 03	Tiles and ceramics.
17 01 07	Mixtures of concrete, bricks, tiles and ceramics other than those mentioned in 17 01 06.*
17 02 01	Untreated wood only.
17 03 02	Bituminous mixtures other than those mentioned in 17 03 01.*
17 05 04	Soil and stones other than those mentioned in 17 05 03.*
17 05 06	Dredging spoil other than those mentioned in 17 05 05.*
17 05 08	Track ballast other than those mentioned in 17 05 07.*
19 05 99	Compost produced pursuant to a treatment described in the paragraphs numbered T23 or T26 of Chapter 2 only.
19 12 05	Glass.
19 12 09	Aggregates only.
19 12 12	Gypsum recovered from construction materials only.
19 13 02	Solid wastes from soil remediation other than those mentioned in 19 13 01.*
19 13 04	Sludges from remediation other than those mentioned in 19 13 03.*
20 02 02	Soil and stones.

** For waste codes beginning with 17 refer to Appendix A, for all other codes refer directly to the European Waste Catalogue.*

E 03

E 03

Appendix F – Waste management, storage and disposal checklist

Company name		Project title	
Location		Contract no.	

Waste management, storage and disposal

	Yes	No	N/A
1. Has a site waste management plan been prepared, identifying waste types and quantities for the project?			
a) Is it being maintained/updated?			
2. Have a waste target and any relevant objectives been set for the project?			
3. Have the project waste procedures and requirements been included in the contracts with sub-contractors and suppliers?			
4. Are the project waste procedures and requirements included in the site induction?			
5. Have waste responsibilities been defined to know who is disposing of what, where and when?			
6. Have designated area(s) been established on site to segregate and reuse waste materials?			
7. Have recycling facilities been established to segregate office or canteen wastes (such as paper, cans and plastics)?			
8. If required, has the site been registered for hazardous waste with the relevant environment agency?			
9. Are waste management permits, licences or exemptions in place for any processing of waste on site (such as crushing and soil treatment)?			
10. Is the WRAP quality protocol being complied with for the production of aggregates from waste?			
11. Is the CL:AIRE Code of Practice being complied with for the treatment and use of contaminated soils (such as appointment of qualified person, materials management plan, and so on)?			
12. Are hazardous wastes (oil, fuel, paints, and so on) collected and stored separately from general wastes?			
13. Is the burning of rubbish on site prohibited, unless a permit or licence has been obtained?			
14. Are registered waste carriers used to remove waste from site and are these checked on a regular basis?			
15. Does the tip where the waste is taken have a licence for the type of waste?			
16. Do waste transfer documents include the right information (six-digit waste codes, licence number of carrier, tip location and declaration regarding the waste hierarchy)?			
17. Are waste transfer documents being retained: two years for non-hazardous and three years for hazardous waste?			
18. Are skips protected and enclosed to avoid waste being blown across site?			

Comments

Name		Position		Signature		Date	

For an interactive PDF version of this form visit the companion website.

04

Energy management

Contents

Supporting information

G: Checklists and forms

GT 700 Toolbox talks

Overview

Global warming and climate change have come to the fore as sustainability issues. How businesses minimise CO_2 emissions is a growing concern for clients, the construction industry as a whole and many other organisations.

This chapter gives a general overview of the UK's legally binding energy and carbon reduction targets. It provides guidance on actions that help construction companies to measure and reduce their energy consumption and associated carbon footprint.

4.1 Introduction

Climate change is now seen as the defining challenge of this era. It is recognised that human activity (through the burning of fossil fuels in energy production, manufacturing and use in buildings and transport) together with the generation of waste create greenhouse gases that contribute to global warming and climate change. Buildings, and the construction activity to produce them, use a significant proportion of this energy and the industry has a major role to play in improving energy efficiency and reducing greenhouse gases and costs.

Action to address the problem of global warming started in 1988, when the intergovernmental panel on climate change (IPPC) was established. The United Nations Earth Summit at Rio in 1992 was, however, the turning point in the fight against climate change.

In December 1997, in Kyoto, Japan, parties to the Climate Treaty (the Kyoto Protocol) agreed to make legally binding cuts in seven greenhouse gases, including carbon dioxide. Between 2008 and 2012 developed countries were to reduce their emissions by an average of 5.2% below 1990 levels. Different targets were set for individual countries and the UK's contribution was a 12.5% reduction. As the Kyoto Protocol expired in 2012 negotiations continued to form the new Paris agreement that came into force in November 2016.

Meanwhile, the UK has implemented the Climate Change Act, which sets out legally binding targets to reduce greenhouse gas emissions by 34% by 2020 and 80% by 2050 based on 1990 levels. In addition, the UK has a target for 15% of all energy to be provided from renewable sources by 2020 from agreement at a European level.

Many mandatory and voluntary initiatives have been introduced by the UK and local governments, as well as the European Union, in response to the risk posed by climate change.

e.g.

Examples include the climate change levy on fossil fuels, revisions to Part L of the Building Regulations, the Energy Performance of Buildings Directive, the EU emissions trading scheme, the renewable heat initiative, the Carbon Reduction Commitment Energy Efficiency Scheme, Energy Savings Opportunity Scheme (ESOS) and the Low Carbon Construction report.

Carbon dioxide accounts for about 82% of all emissions from the seven main greenhouse gases. Almost half of these emissions are associated with buildings. In the UK this amounted to approximately 465 million tonnes in 2010, dropping to 381 million tonnes in 2016, mainly as a result of the reduction in fossil fuel powered electricity generation.

The Green Construction Board has developed the Low Carbon Routemap and action plan for the Built Environment, enabling customers to understand the policies, actions and key decision points required to achieve the UK Government target of 80% reduction in greenhouse gas emissions in the built environment, against 1990 levels, by 2050. The routemap also sets out actions, together with key performance indicators, that can be used to deliver and measure progress in meeting the 2050 target.

www.

For further details of the Low Carbon Routemap and action plan refer to the Green Construction Board website.

The UK Government's *Construction 2025: industrial strategy for construction*, published in July 2013, sets out an aspiration for a 50% reduction in greenhouse gas emissions in the built environment. This includes an action plan to develop market and technology-based plans to secure the jobs and growth opportunities from driving carbon out of the built environment, led by the Green Construction Board.

www.

For further details about *Construction 2025* refer to the Government website.

UK carbon pollution is at its lowest level since 1894, apart from during the coal mining disputes in the 1920s.

4.2 Important points

Energy is a valuable resource. Reducing the amount used and developing more efficient methods of generating energy can provide significant savings and reduce the emission of greenhouse gases, which contribute to climate change.

A construction project can save energy in two main ways. First, by managing the energy used for the construction phase and second, by constructing the project in a way that ensures the completed structure gains its optimum energy performance for its lifetime use. *(For further information refer to 4.5.4.1 Quality and workmanship later in this chapter.)*

The following site practices should be employed to reduce energy consumption and greenhouse gas emissions.

- ☑ Engage with distribution network operators and energy suppliers as early as possible to obtain a connection to the site to avoid the use of petrol or diesel generators.
- ☑ Ensure that all plant and equipment is switched off when not required. Reinforce through monitoring and inspections.
- ☑ Ensure that all plant and equipment is regularly maintained.
- ☑ Put an efficient transport logistics plan in place, to reduce the need for lorry movements and double handling.
- ☑ Consider energy efficiency technologies for site accommodation (for example, LED lighting, PIR sensors and double glazing).
- ☑ Monitor and regularly report site energy and fuel use, and positively reinforce reductions through awareness and toolbox talk sessions.

Construction detailing and quality of workmanship on site can potentially have a major impact on the performance of the building in use. Gaps in insulation at junctions between windows and doors, for example, can cause cold bridging leading to reduced air tightness, increased heat loss and subsequent energy use.

Building performance can also be optimised through effective handover and commissioning processes called soft landings. Follow up and seasonal commissioning, including the appropriate training of operational staff to use the building's control systems, can also significantly improve the performance of a building in use.

4.3 Managing energy on site

4.3.1 Using and conserving energy

Using less energy during any construction activity is an ideal way to cut costs and save money. In response to the 15% reduction target set out in the previous *Strategy for sustainable construction*, the Strategic Forum for Construction (SFfC) and the Carbon Trust prepared an action plan that identifies what steps can be taken to deliver this target. The main areas for action include on-site construction (energy use, plant and equipment) and site accommodation, transport associated with the delivery of materials and removal of waste, business travel and corporate offices.

The action plan sets out a number of measures that can be undertaken by companies to help reduce their energy use and carbon emissions.

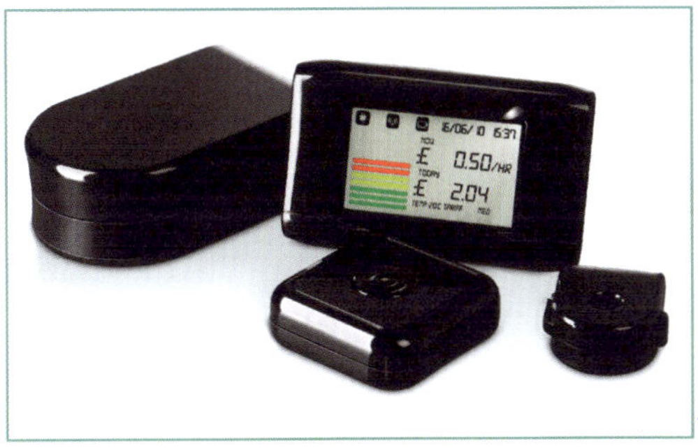

Meter for monitoring electricity use and cost

- ☑ Ensuring sites connect to the electricity supply as early as possible to prevent lots of equipment running on fuel-driven generators.
- ☑ Using alternative or recycled products for site set up, logistics and enabling works.
- ☑ Using products with a lower carbon footprint.
- ☑ Installing more energy-efficient site accommodation (for example, environmental cabins containing such items as PIR sensors on lights, double glazing, additional insulation, door closers, waterless urinals and fuel cell and solar panel technology).
- ☑ Efficient use of construction plant through induction and training (such as turning off when not required and keeping them well maintained).
- ☑ Good practice energy management on site (for example, festoon low-energy light bulbs and turning task lighting off when not in use).
- ☑ Consolidation areas and sharing of transport: the use of off-site consolidation areas and the sharing of transport will reduce the number of vehicle movements to and from site. Vehicle sharing can be planned by site staff. Project suppliers may permit the use of their parking areas as car-pooling points.
- ☑ Fuel-efficient driving through driver training.
- ☑ Good practice energy management in company offices.

Whilst the action plan focuses on the direct use of energy, significant savings in embodied energy can also be achieved by using low energy products (such as recycled aggregates).

www. **For good examples of reducing energy on sites refer to the *Infrastructure carbon review*.**

4.3.2 Transport

Vehicle movements and the transport of materials and people represent energy resources being used. It is important that vehicle movements are handled, programmed and managed in an efficient way, as this will lead to greater resource efficiency and reduced costs.

Good logistics, with prompt arrivals and departures and materials being off-loaded in the right locations, will result in multiple positive benefits. Perhaps the key amongst these will be reduced fuel use and therefore lower carbon emissions, which brings associated cost savings. Crucially there is also less potential for double handling and wastage of materials.

Efficient vehicle logistics will also lead to reductions in several negative impacts. Efficient logistics with vehicles arriving promptly, being quickly off-loaded, with fewer delays and departing quickly, mean that several types of potential impact will be reduced.

- ☑ There will be less time when noise is being generated. Effective vehicle sequencing will also prevent traffic build-up noise being generated from multiple sources.
- ☑ There will be less time that vehicles are on site, with fewer vehicles sitting with idling engines, resulting in lower exhaust emissions. This will reduce the potential adverse impact on air quality.
- ☑ Vehicle arrivals and departures that are sequenced in the right way will have less of an impact on local traffic flows and reduce local congestion.
- ☑ Effective traffic management will result in less dust and mud being produced. This will reduce the amount of road sweeping and dampening down having to be incorporated into the traffic management plan.

www. **For further information on business efficiency visit the Lean Enterprise Institute website.**

The flow of vehicles, materials and resources onto a site needs to be properly sequenced and planned. This process needs to be undertaken with a proper understanding of local traffic flows and factors that will influence the logistics plan (such as school run times). A lack of understanding will result in congestion, delays in material delivery and a variety of negative impacts occurring.

An effective traffic management plan will lead to improved public relations with the surrounding community and businesses. Traffic to and from a site, and flow of materials and people with that traffic, are one of the main areas of interaction with a location, and it is important to communicate these arrangements effectively. The right interface with, and respect for, a location should lead to trust and appreciation in return from the people in that area.

A good traffic management plan will give consideration to the following factors

- ☑ Local traffic conditions, peak flows and congestion hot spots.
- ☑ Delivery and departure sequencing and times for all site vehicles.
- ☑ Off-site delivery and departure routes for all site vehicles.
- ☑ Signage and directions on site.
- ☑ Hold areas for vehicles waiting to off-load or depart.
- ☑ Communication between vehicles and the logistics manager on and off site.
- ☑ Wheel-wash locations, dust suppression and mud sweeping.
- ☑ Receiving and responding to complaints.
- ☑ Consultation processes with emergency services, local residents, schools, public facilities and businesses.
- ☑ Consolidation areas and sharing of transport: the use of off-site consolidation areas and the sharing of transport will reduce the number of vehicle movements to and from site. Vehicle sharing can be planned by site staff. Project suppliers may permit the use of their parking areas as car-pooling points.
- ☑ Parking and lay down areas:
  - on-site parking for staff, contractors and visitors needs to be clearly identified
  - delivery vehicle and waste removal parking also needs identifying
  - vehicle off-loading and lay down areas should be clearly identified.
- ☑ Ensure that plant and vehicles are properly maintained to ensure efficiency of operation.
- ☑ On-site traffic management (safe routes, dust and wheel wash):
  - use wheel washes to prevent mud and any contaminated materials getting onto local roads
  - water bowsers can be used for dust suppression
  - road sweepers can be used to remove any mud accumulations.
- ☑ Promotion of opportunities and arrangements for green travel for site personnel and visitors.

For energy and transport checklists refer to Appendices B and C.

4.4 Measuring energy and carbon

Measuring overall energy usage and being able to identity where energy is being used and in what form is a prerequisite to starting an energy and carbon management programme. To do so will require the use of meters to measure energy usage, and on large sites this may involve sub-meters. Sites may be required to measure journeys to and from the project associated with deliveries and business travel and measure the embodied energy associated for the materials they are using.

Embodied energy is defined as the commercial energy (for example, fossil fuels and nuclear) that was used in the processes to make any product, bring it to market and dispose of it. Embodied energy is an accounting methodology that aims to find the sum total of the energy necessary for an entire product life cycle. This life cycle includes raw material extraction, transport, manufacture, assembly, installation, disassembly, deconstruction and/or decomposition.

To measure the carbon footprint associated with energy use will require the use of conversion factors for each energy type. Different methodologies produce different understandings of the scale and scope of application and the type of energy embodied. Some methodologies account for the energy embodied in terms of the oil that supports economic processes. There are three levels of assessment that are currently being used.

- **Scope 1** are the emissions from sources under the immediate control of the company.
- **Scope 2** are the off-site emissions from the purchase of electricity.
- **Scope 3** are the off-site emissions from the company's supply chain or from products sold by the company.

The Greenhouse Gas Protocol developed by the World Business Council for Sustainable Development (WBCSD) and the World Resources Institute (WRI) is the global standard for the measurement and reporting of Scope 1, 2 and 3 carbon emissions.

The Environment Agency has a carbon calculator for construction work, which is hosted on the Government services and information website. This is an Excel spreadsheet that calculates the embodied carbon dioxide of materials plus carbon dioxide associated with their transportation. It also considers personal travel, site energy use and waste management. The tool has a number of benefits, some of which are shown below.

- Helping to assess and compare the sustainability of different designs, in carbon dioxide (CO_2) terms, and influences option choice at the options appraisal stage.
- Helping to highlight where big carbon savings on specific construction projects can be made.
- Calculating the total carbon footprint from construction and helping to reduce it.

The tool was developed by the Environment Agency for its own construction work (predominantly fluvial and coastal construction projects). However, other construction clients, contractors and consultants may find it useful when assessing their own work.

For carbon reduction case studies refer to Appendix A.

Measuring and reporting of site energy use is also a requirement under the management credits of BREEAM. At present, BREEAM sets no requirement on the use of a particular method or protocol for reporting energy/carbon from construction sites, as there are currently no uniformly accepted protocols for the collection of data and assessment of emissions from construction sites. (*Source: Carbon: Reducing the footprint of the construction process, an action plan to reduce carbon.* Strategic Forum and Carbon Trust, 2010.)

The CEEQUAL assessment scheme requires both energy and carbon assessments.

The European Network of Construction Companies for Research and Development (ENCORD) has established a *Carbon measurement protocol*. The document identifies the intended users of the protocol, the main sources of emissions over which a construction company may have some influence, and the method of measuring these emissions. Guidance is also provided on reporting methods at a company and project level, with a view that companies will report their emissions publicly. This is also intended to assist current and future work undertaken to reduce emissions from specific construction-related work and operations.

As the energy supply becomes greener, the proportion of energy locked up in building materials increases. The greatest carbon use in the life cycle of a building is during use. Therefore, designers need to think about both embodied and operational energy of the building.

4.4.1 Carbon Reduction Commitment Energy Efficiency Scheme

The Climate Change Act introduced new regulations from 1 April 2010, through the Carbon Reduction Commitment (CRC) energy efficiency scheme and amending Order in 2013, to improve the energy efficiency of larger companies. It applies to England, Northern Ireland, Scotland and Wales. These regulations require companies to measure and report their carbon emissions and to purchase carbon allowances in line with their carbon footprint. Some large construction companies may fall under these requirements.

The CRC energy efficiency scheme is being run in phases. Phase 1 started in April 2010, Phase 2 started in April 2013 (registration year) and subsequent phases will start in April 2019. The Environment Agency is the administrator for the CRC energy efficiency scheme.

All companies that purchased electricity through mandatory half-hourly meters from 1 April 2012 to 31 March 2013, and had a total consumption of more than 6,000 MWh through these meters, were required to participate in Phase 2 of the CRC energy efficiency scheme, by completing the following.

- ☑ Registering the company by 31 January 2014 for Phase 2 of the scheme by providing details of its structure, meters, relevant contacts and paying the appropriate registration fee.
- ☑ From 1 April 2014, recording all details of suppliers and consumption for electricity and gas (excluding transport).
- ☑ Maintaining a formal evidence pack that includes details of all qualification, supply and consumption data, together with details of company changes, CRC responsibilities, and so on.
- ☑ Carrying out an internal audit and certifying the evidence pack.
- ☑ Calculating the carbon footprint of the company from the use of its qualifying electricity and gas (excluding transport).
- ☑ Purchasing relevant carbon allowances each year. In Phase 2 of the scheme companies can purchase carbon allowances, equal to their estimated carbon footprint, in a forecast sale as well as a 'buy to comply' sale. The cost allowances in the forecast sale are slightly cheaper, allowing companies who can accurately estimate their emissions to save money in carbon taxes.
- ☑ Submitting a formal annual report to the regulator in July of each year.
- ☑ Participating in audits by the scheme administrator when requested.

E 04

In the budget on 16 March 2016, the Chancellor of the Exchequer announced that the Government had decided to close the CRC scheme following the 2018-19 compliance year.

Organisations will report under the CRC until the end of July 2019 and surrender allowances for emissions from energy supplied in the 2018-19 compliance year by the end of October 2019.

www.

For comprehensive guidance on the CRC *energy efficiency scheme: qualification and registration* refer to the Government website.

4.4.2 Energy Savings Opportunity Scheme (ESOS) Regulations

The Energy Savings Opportunity Scheme (ESOS) is a mandatory energy assessment scheme for organisations in the UK that meet the qualification criteria of being a large undertaking, in line with one or both of the following conditions.

- ☑ Employing 250 or more people.
- ☑ Having an annual turnover in excess of 50 million euros (£38,937,777) and an annual balance sheet total in excess of 43 million euros (£33,486,489).

The Environment Agency is the UK scheme administrator.

ESOS applies to large UK undertakings and their corporate groups. It mainly affects businesses but can also apply to not for profit bodies and any other non-public sector undertakings that are large enough to meet the criteria.

Organisations that qualify for ESOS must carry out ESOS assessments every four years. These assessments are audits of the energy used by their buildings, industrial processes and transport to identify cost effective energy saving measures.

Qualifying organisations must notify the Environment Agency by a set deadline that they have complied with their ESOS obligations.

Various penalties apply for a breach of these regulations. In particular, in relation to providing a false or misleading statement, and also in relation to a failure to:

- ☑ notify
- ☑ maintain records
- ☑ carry out an energy audit
- ☑ comply with a notice.

www.

For further details of the ESOS scheme and qualification criteria refer to the Department of Energy and Climate Change.

4.5 Delivering energy performance in buildings

When considering energy use and emission reduction, this chapter has so far focused on construction site activities. Equally, there are important contributions that can be made to ensure completed buildings are energy efficient: warm in winter and cool in summer, with excellent indoor air quality, low energy bills and low carbon emissions. Every building will consume energy and create emissions throughout its operational life. It is therefore important that buildings and projects are completed to ensure optimum energy performance and that clients help to reduce the costs and carbon emissions associated with this energy use.

4.5.1 Achieving energy efficiency

Buildings should work as an energy system in which heat gains and losses are always in balance. The higher the loss from the building fabric, the more energy is needed to heat or cool it. Buildings lose heat through the external envelope in two ways: firstly, by a complex mix of conduction, convection and radiation through the materials and air spaces in the construction, and secondly via direct air leakage from inside to outside. This means that the design of the building envelope is critical.

Insulation layers and the associated detailing at junctions need to be designed and constructed with considerable care, based on a good understanding of the heat loss mechanisms involved. In addition, plant and equipment need to be correctly sized, energy-efficient and the whole service system designed, installed, commissioned and operated to achieve maximum overall efficiency.

4.5.2 Closing the performance gap

Recent research has revealed a huge gap between how much energy a building is designed or predicted to use and its actual energy use, which is often much higher – in some cases as much as five times the designed or predicted energy consumption.

Although some of this can be attributable to variations in use, which is difficult to predict, a large element is related to the technical under-performance of fabric and services due to design and construction failings; this is referred to as the *performance gap*. One of the important contributors to the performance gap is a lack of understanding of energy efficiency within the built environment workforce.

It is now common for building performance to be either measured during construction (for example, air tightness tests) or after the building is complete, to identify how it performs whilst in use, rather than just relying on a statement of the design intent.

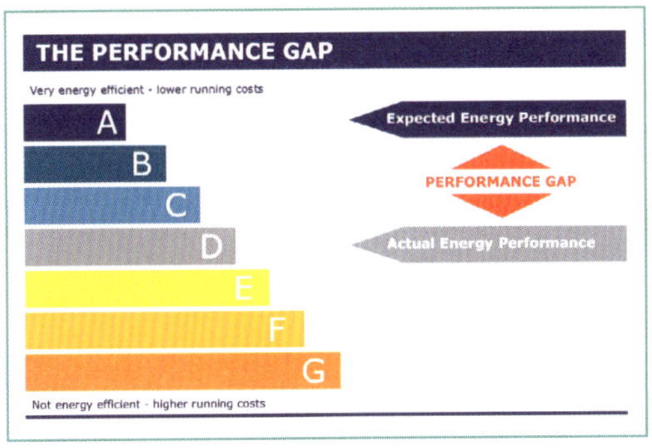

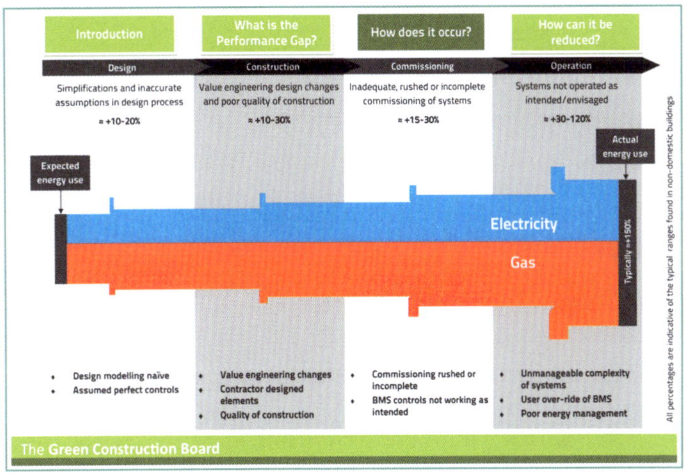

Closing the performance gap (Images supplied by the Green Construction Board)

For further information and a copy of the *Sustainable building training guide* visit the Construction Leadership Council's website.

4.5.3 Understanding performance

Measuring buildings by how well they perform is carried out against Building Regulations Part L, their energy performance certificates (EPCs) and other certification processes (such as BREEAM, LEED and CEEQUAL). All of these systems indicate how the energy efficiency of the design and construction of the project should perform **on paper**. Understanding how buildings are actually performing during occupation or operation is the first step in the process to improving this.

An important factor is how the regulations check how sustainable designs and performance promises work out **in reality**. There are many ways that this can be measured for buildings, but one of the most common in the UK is the display energy certificate (DEC). The DEC evaluates how much energy the building actually uses during 12 months of occupation, and compares this performance with the typical use for buildings of a similar type.

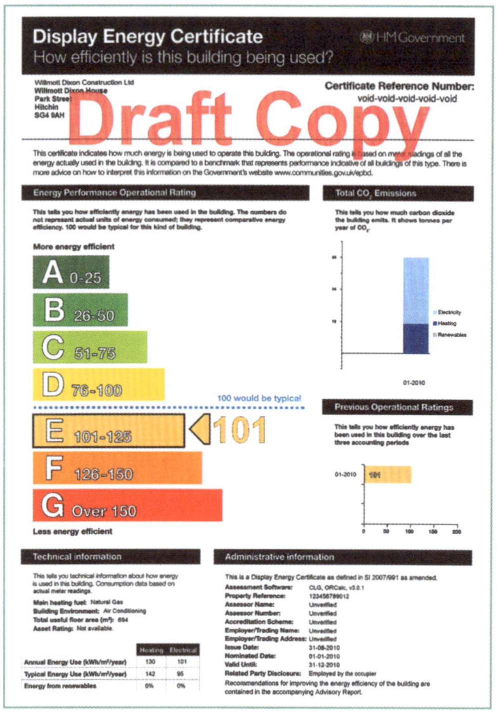

Display Energy Certificate
How efficiently is this building being used?
HM Government

Willmott Dixon Construction Ltd
Willmott Dixon House
Park Street
Hitchin
SG4 9AH

Certificate Reference Number:
void-void-void-void-void

Draft Copy

This certificate indicates how much energy is being used to operate this building. The operational rating is based on meter readings of all the energy actually used in the building. It is compared to a benchmark that represents performance indicative of all buildings of this type. There is more advice on how to interpret this information on the Government's website www.communities.gov.uk/epbd.

Energy Performance Operational Rating

This tells you how efficiently energy has been used in the building. The numbers do not represent actual units of energy consumed; they represent comparative energy efficiency. 100 would be typical for this kind of building.

More energy efficient
A 0-25
B 26-50
C 51-75
D 76-100
100 would be typical
E 101-125 — 101
F 126-150
G Over 150
Less energy efficient

Total CO_2 Emissions

This tells you how much carbon dioxide the building emits. It shows tonnes per year of CO_2.

Electricity
Heating
Renewables
01-2010

Previous Operational Ratings

This tells you how efficiently energy has been used in this building over the last three accounting periods

01-2010 — 101
0 50 100 150 200

Technical information

This tells you technical information about how energy is used in this building. Consumption data based on actual meter readings.

Main heating fuel: Natural Gas
Building Environment: Air Conditioning
Total useful floor area (m^2): 694
Asset Rating: Not available.

	Heating	Electrical
Annual Energy Use ($kWh/m^2/year$)	130	101
Typical Energy Use ($kWh/m^2/year$)	142	95
Energy from renewables	0%	0%

Administrative information

This is a Display Energy Certificate as defined in SI 2007/991 as amended.

Assessment Software: CLG, ORCalc, v3.0.1
Property Reference: 123456789012
Assessor Name: Unverified
Assessor Number: Unverified
Accreditation Scheme: Unverified
Employer/Trading Name: Unverified
Employer/Trading Address: Unverified
Issue Date: 31-08-2010
Nominated Date: 01-01-2010
Valid Until: 31-12-2010
Related Party Disclosure: Employed by the occupier

Recommendations for improving the energy efficiency of the building are contained in the accompanying Advisory Report.

Energy certificate (Image supplied by Willmott Dixon)

The way that projects are actually performing is not always known to the teams building them, creating a risk of them becoming disconnected. Research has shown that actual performance differs from clients' expectations of the regulation levels by a significant degree. This is known as the **performance gap**.

Industry bodies (such as the UK Green Building Council and Zero Carbon Hub) have published numerous case studies explaining these issues in detail. The important point for the industry is to know how it can act to improve the energy performance of a building at design, construction, commissioning and handover stages of the project.

4.5.4 Improving performance

4.5.4.1 Quality and workmanship

As building regulations and air tightness standards for new and refurbished buildings become more demanding, the quality of materials and the workmanship, skills, knowledge, training and experience of the workforce required to install them has a significant impact on energy efficiency. It is therefore important that all site personnel understand how energy-efficient construction can be achieved, to deliver good energy performance in reality.

An example of poorly fitted cavity wall insulation board (Image supplied by Willmott Dixon)

Incorrect detailing and substituting materials for inferior performing alternatives (which may look the same) result in buildings that will perform poorly throughout their lifespan. In turn, this will result in a significant increase in energy use and running costs.

Insulation quality. Poorly fitted insulation causes cold bridging and heat loss. This can significantly reduce the energy performance of a building.

Cavity wall insulation board, which has been poorly fitted, has been measured as creating a 300% greater heat loss than designed values.

Testing for air tightness (Image supplied by Willmott Dixon)

Air tightness. Air gaps, caused by poor detailing, unsealed service penetrations and poorly fitted doors and windows, allow more air infiltration than the designed values. These gaps will increase heat loss or heat gain, leading to a higher demand for additional energy for heating or cooling.

The lower the air tightness testing result in $m^3/(hr.m^2)$, the lower the unintended heat loss or gain will be. Performance, greater than 10 times the Building Regulations minimum standard, has been achieved on many UK construction projects.

E 04

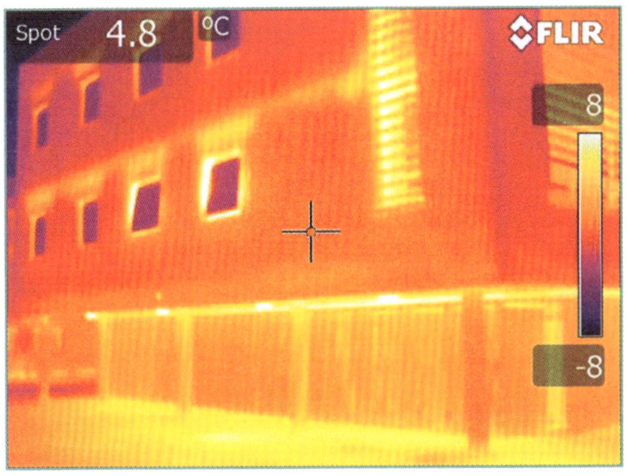

Thermal image showing excessive heat loss above second floor window heads, indicative of poor continuity of thermal insulation (Image supplied by Willmott Dixon)

Cold bridging. Significant heat loss can be caused through cold bridges, produced by poor design, changes in the design details or poor workmanship. This heat loss can create condensation problems and damp when the building is in use. Installing insulation to provide a continuous barrier, particularly at junctions of walls, roofs, floors and around windows, is important in avoiding this.

4.5.4.2 Low carbon renewable energy

As part of the UK's target to reduce carbon emissions, 15% of all energy must come from renewable sources by 2020. The following types of low carbon renewable energy are typically used on construction projects.

Photovoltaic power

☑ **Renewable electricity.**

- Solar photovoltaic (PV).
- Wind.
- Hydro-electricity, including tidal energy.
- Electricity generated from anaerobic digestion (AD).
- Electricity generated from combined heat and power (CHP).

Biomass heating

☑ **Renewable heat.**

- Biomass heating.
- Solar thermal hot water.
- Ground source heat pumps.
- Air source heat pumps.
- Heating generated from biomass-fuelled CHP.

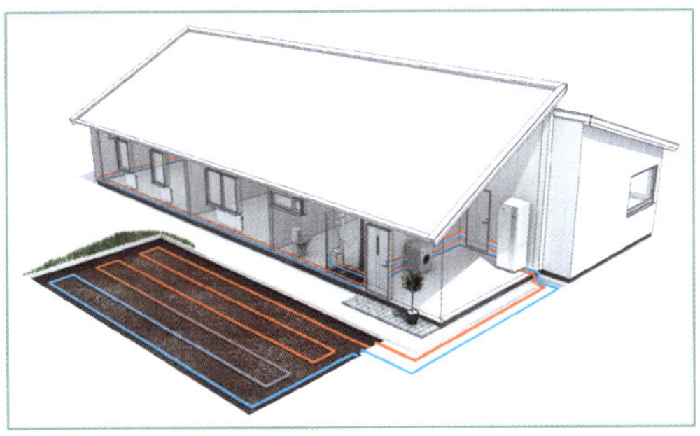

Ground source heat pump

The UK Government gives financial support through feed in tariffs (FITs) for renewable electricity technologies, and renewable heat incentive (RHI) for renewable heat technologies. These incentives are primarily designed to stimulate take-up of these technologies until the market is sufficiently developed and prices fall.

Photovoltaic power, for example, has seen a rapid increase in deployment over the last few years because of the generous FITs that have been paid. As the market has been established these tariffs are reducing. In addition, because renewable energy reduces the amount of fossil fuels supplied but not the demand (energy efficiency), it is now a requirement to demonstrate that the building has achieved a certain level of efficiency before full FITs are given.

Renewable energy reduces the overall carbon emissions of the building or project. However, the correct way to use the technology is to first consider energy reduction approaches (such as more insulation, better performing windows or glazing, or more efficient building services). Renewable energy can then be used to further reduce emissions and costs from the remaining energy needed. Renewable energy will, however, be a vital element of achieving zero carbon energy requirements set out in standards such as the Building Regulations and BREEAM, together with achieving zero carbon in non-domestic buildings by 2018.

4.5.4.3 Commissioning, handover and aftercare

Making sure that a building and its mechanical and electrical services are working properly and are handed over to the client in an effective way are vital steps to ensure it operates as it was intended to, and that energy consumption in use is minimised. The process, known as soft landings, is a step-by-step approach to ensuring that this is achieved as part of the project.

Some important steps in the soft landings approach are shown below.

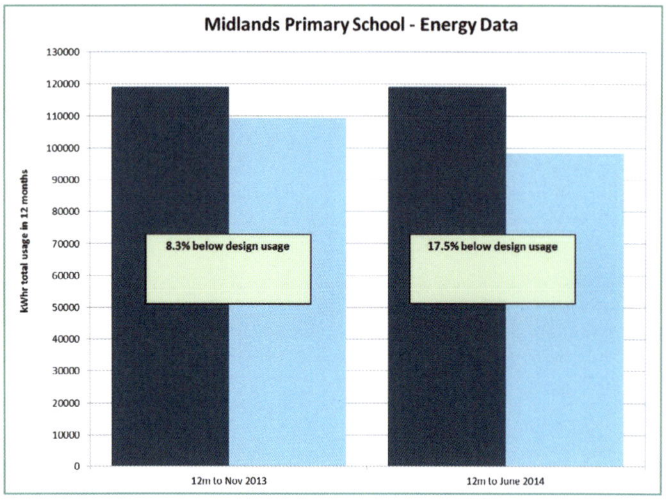

Energy efficiency improvements following POE actions (Image supplied by Willmott Dixon)

- **Developing realistic targets for energy consumption in use.** This is achieved by building on the Building Regulations' calculations and then adding additional estimated usage for equipment such as IT systems, kitchens, external lighting and other use that is not counted in the Building Regulations Part L calculations. In addition, there should be an allowance for realistic usage conditions; knowing 'what the building should be achieving' is the first step in judging its success.
- **Ensuring that energy meters measure what the targets are set up to assess.** This makes monitoring for the client much easier and costs no more than usual.
- **Fully commissioning and testing prior to handover.** This process is often under pressure, due to complete and occupy, but time and effort invested is well spent. Ensuring that follow-up recommissioning or seasonal commissioning takes places has been shown to make a significant difference to actual usage.
- **Supplying simple, effective user guides and comprehensive training for building staff.** This supports low energy use in operation and reduces complications from follow-up demand for site support after handover.
- **Aftercare processes.** As well as resolving defects in the liability period, these also focus on helping the client to manage their new building effectively. Building management systems are often complex, and offers of additional client training in their use can be cost effective.
- **Measuring and monitoring actual performance in use.** This is sometimes called *post occupancy evaluation* (POE). It can often identify where energy efficiency improvements can be made, saving both energy and costs.

For further details of the soft landings approach and free guidance refer to the BSRIA website.

UK construction is already delivering quality projects every day, on time and on budget. Focusing on these issues will make 'on performance' the next major achievement.

4.5.5 Builders' book

An informative guide has been produced for the house building sector, which promotes good craftsmanship and highlights important construction details when building new homes. Aimed at on site personnel, the Zero Carbon Hub *Builders' book* recommends solutions to the most common construction issues that may lead to poor energy performance, loss of comfort and increased energy bills in new build homes.

The bulk of the book is made up of site posters that highlight areas of the build in a normal construction sequence. These posters can be printed out in A4 or A3 size, in colour, and can be used as site guidance for site management and trades. They can be used to inform toolbox talks for trades and at site inductions, as well as design documents for inclusion in specification documentation.

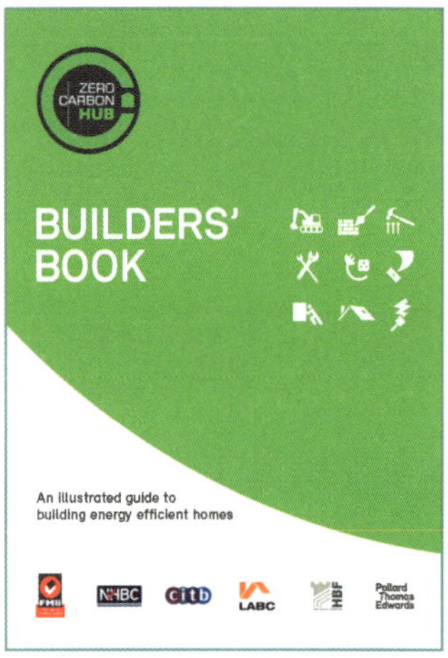

Zero Carbon Hub Builders' book

4.5.5.1 Examples of Builders' book contents

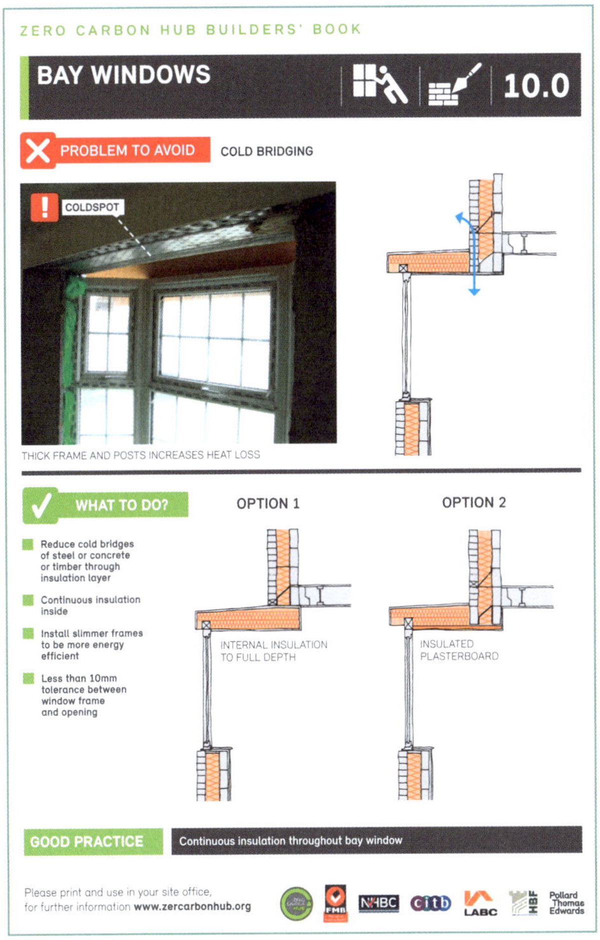

Reducing cold bridges through insulation layer

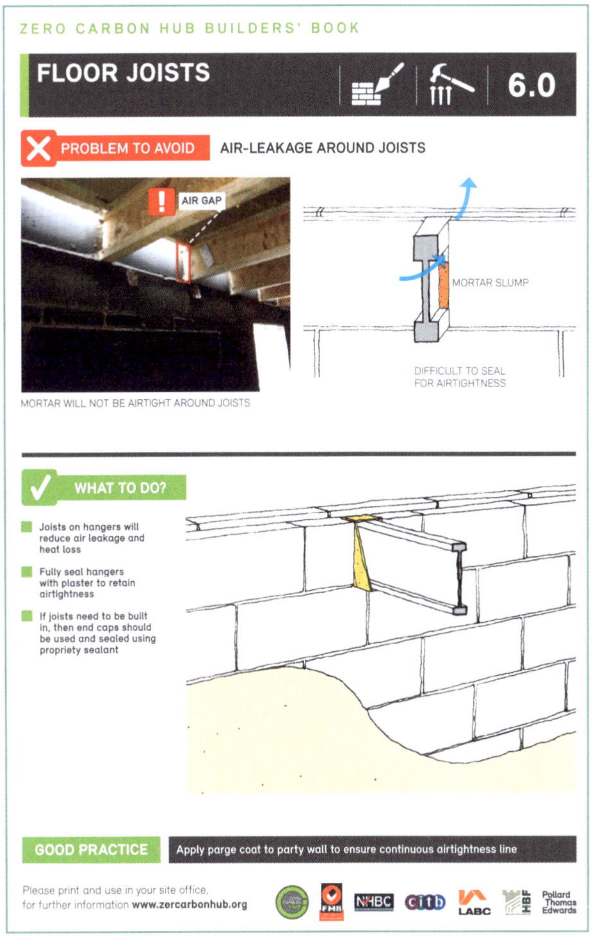

Avoiding air leakage around joists

The Zero Carbon Hub ceased operations on 31 March 2016 as a result of withdrawal of industry funding. The website and associated guidance documents are still currently accessible.

For further information refer to the Zero Carbon Hub website.

E 04

Appendix A – Carbon reduction case studies

To access the *Construction carbon calculator*, referred to in some of the case studies below, refer to the Government website.

Sandford bridge

Background. The Sandford bridge project (£380,000) spans the Sandford lock bypass channel on the River Thames. The bridge is primarily used for access to the lock house and the lock structure, but additionally there is a 3 m wide roadway spanning 40 m across the river. The aim of the bridge refurbishment was to increase the carrying capacity from 3 to 18 tonnes.

Reducing the carbon footprint. The design team were interested in minimising the environmental impact and maximising carbon savings throughout the duration of the project. Overall the team saved 62 tonnes of CO_2 (from 125 tonnes CO_2 to 63 tonnes CO_2).

The ideal solution was to reuse the existing sub-structure, replace the old deck with a new one and construct a new vehicle restrain parapet. The original bridge deck, constructed from large, longitudinal concrete slabs, was removed to install a new deck made from a steel frame and comparatively smaller pre-cast concrete slabs. The benefit of the latter was the requirement for a smaller crane for lifting and positioning, which reduced the carbon emissions from 48 tonnes of CO_2 to 8 tonnes of CO_2.

The old deck material was crushed on site then reused to reinstate the car park, saving 2 tonnes of CO_2 by avoiding waste off site and import of aggregates.

The design of the vehicle parapet height was reduced from 1.5 m to 1 m. This was mainly for aesthetic purposes but it saved 2 tonnes of CO_2 from the amount of steel required and transportation.

Instead of using a sheet pile wall to separate the bank from the river the existing scour pile in the right-bank of the bridge was reused, instead of removing it, saving 18 tonnes of CO_2 in materials and transport.

The learning curve. The Sandford access bridge case study provides an example of a project seeking to maximise the reuse of existing site materials, which has led to a reduction in embodied CO_2 emissions, as well as making wider environmental gains and cost savings.

(Reproduced with permission from the Environment Agency – Case study IEM/2012/002.)

Weybridge 24-hour moorings

Background. Weybridge 24-hour moorings in Surrey was a £375,000 project to build a 120 m long river level footpath with access to moorings. The original design was based on the construction of an in-situ concrete wall supported by steel sheet piling. However, a design review was undertaken (primarily due to cost constraints and to meet a completion date). The design was changed to utilise less costly and (overall) lower carbon materials whilst maintaining the same operational life.

Reducing the carbon footprint. The post construction emission was reduced by 169 tonnes of CO_2 compared with the original design (from 255 tonnes of CO_2 to 86 tonnes of CO_2).

Mesh filled concrete. The most significant CO_2 saving was from the 75% reduction in concrete used in the wall. A mesh filled with concrete was used instead of precast concrete blocks. This saved 173 tonnes of CO_2 (59 tonnes of CO_2 compared to 232 tonnes of CO_2 for precast concrete blocks). Originally a cast in-situ concrete wall was planned, which involved considerably more material.

Use of plastic piles. Further carbon savings were achieved through the innovative use of plastic piles (89.5% recycled) instead of steel sheet piles. The carbon footprint of the plastic piling was determined to be 8.6 tonnes of CO_2 compared to 17 tonnes of CO_2 for steel sheet piles.

Weybridge 24-hour moorings (continued)

Form liner. Using a dense foam form liner saved considerable time and cost compared to an authentic alternative. A brickwork finish was achieved to fit the surroundings using a reusable rubber form liner. This added 4 tonnes of CO_2 to the project.

Concrete specification. The concrete specification/grade was changed from exposure class XC3 to XC4 to increase the speed of the construction to meet a completion deadline. The carbon footprint increased by 27 tonnes, from 228 tonnes to 255 tonnes of CO_2 due to the time constraints.

The learning curve. The project was delivered on time and under budget. Overall the project team saved approximately £40,000 compared to the original design, with the same design life. Although the drivers for the fresh look at the design and consideration of innovative materials and approaches were predominantly cost and time, the carbon footprint reduced overall by 50%. This shows that innovation can have multiple benefits. It is well worth having a fresh look even without constraints being imposed.

(Reproduced with permission from the Environment Agency – Case study IEM/2012/001.)

Nottingham left bank

Background. Nottingham left bank is a £51 million scheme that will protect 16,000 homes and businesses from flooding. The project spans a 27 km reach of the River Trent and comprises flood defences with a mixture of earth bunds, concrete walls and sheet piling.

Due to constraints from working next to a railway and nature reserve, and the requirement to prevent groundwater seepage through the underlying sands and gravels, the project team came up with an innovative method that also lowered the total project carbon footprint.

Reducing the carbon footprint. Using the *Construction carbon calculator,* the total carbon footprint of the Nottingham project was 11,800 tonnes. The most significant contributions were from steel, concrete and material transport. The project team implemented many innovative changes to reduce the carbon footprint of the project. One of these examples resulted in changing the material and method of construction, which saved approximately 2,500 tonnes of CO_2 during this section of the works.

Use of TrenchMix. Sheet piles are traditionally used to form a cut-off to protect against groundwater seepage through defences, but on this project a large section of piling was replaced by the use of TrenchMix.

This process involves the mixing of the soil with cement; this cementitious material then sets forming a barrier with a greatly reduced permeability. The soil and cement are mixed by a modified drainage trenching machine (*see photo*).

By using this method the team reduced the amount of steel used on the project by two thirds, saving around 1,876 tonnes of CO_2.

Another benefit of the reduction in steel is the reduced carbon cost of transportation, saving around 700 tonnes of CO_2. The raw materials (cement) and plant emissions from the operation have a footprint of 330 tonnes of CO_2.

The learning curve. Even when taking into account the footprint of the materials and emissions from the TrenchMix method, it is still carbon beneficial by about 70% compared to the sheet pile option. Other advantages include less noise and vibration than sheet piling. This could be useful when working near sensitive areas, such as nature reserves and residential properties.

(Reproduced with permission from the Environment Agency – Case study IEM01/2011/002.)

Burrowbridge bank

Background. The Burrowbridge bank is a £270,000 project to repair a failing asset. This bank is part of the Parrett tidal reach and had suffered erosion, lost many of its timber piles that provided protection and had been affected by landslides. The project trimmed the riverside slope to a more suitable profile while maintaining crest width by filling at the rear. The embankment's toe was also protected by new timber piles and stone while a soft engineering solution was used for the bank slope.

Reducing the carbon footprint. Using the *Construction carbon calculator,* the predicted total carbon footprint of the Burrowbridge project was 140 tonnes. The most significant contributions are from import of clay and timber materials to site and export of trimmed arisings off site.

Reuse of material. The project team challenged the specification for re-profiling of the rear slope of the bund. This meant that material trimmed from the riverside slope could be reused on the land side of the bank. This saved having to import 550 tonnes of clay, reducing the number of lorry movements by 30 and saving around 15 tonnes of CO_2. It also reduced the transport of similar volumes of arisings off site, while providing cost savings of £5,500.

Reducing material used. Access to the site was challenging. Instead of using a traditional stone access track, the team hired in a temporary trackway, which allowed the team to go past the Burrow Mump (scheduled ancient monument) with minimum disruption. Using this trackway saved the need to import 1,500 tonnes of quarried material, which saved 20 tonnes of CO_2 and reduced an additional 100 lorry movements.

Recycled materials. Recycled hardwood timber piles were sourced from a local wharf being demolished, which saved approximately £35,000 and 26 tonnes of CO_2 (compared to using virgin hardwood).

The learning curve. This relatively small project reduced its total footprint by 60 tonnes of CO_2 – a reduction of over 40% of the predicted footprint. This project team proved that by looking at material selection and challenging the specification, large savings can be made not only in carbon and project costs.

(Reproduced with permission from the Environment Agency.)

Radcot weir

Background. The Radcot weir project is one of five sites in the Paddle and Rymer package. The main construction work at Radcot involves removing the existing Paddle and Rymer structure and replacing it with a dipping radial gated weir. This will benefit operators of the weir through the removal of health and safety risk, easier operation and standardisation of the weir. In addition the project will provide a new bypass channel to allow upstream and downstream movement of fish and provide recreational use for canoeists.

Using the *Construction carbon calculator*, the total carbon footprint of the Radcot weir project is 600 tonnes. The most significant contributions are from concrete and steel. The project team, by looking at material selection, have saved around 50 tonnes of CO_2 during the first phase of works.

Reducing the carbon footprint. Use of granular ground blast furnace slag (ggbs) in concrete; 50% ggbs replacement was used in the base and 70% in the rest of the structure. This saved around 40 tonnes of CO_2 (a saving of over 60% when compared to a CEM1 concrete).

Reuse of material. The old structure was demolished and then crushed on site. This material was then used underneath the blinding instead of a primary aggregate, saving both the cost and carbon of disposal of the material and the import of primary aggregate. This saved around 5.2 tonnes of CO_2, resulted in 40 less lorry movements and saved £10,000.

Avoiding waste. The team cast the coping stones for the structure on site by using the leftover concrete from the back end of the concrete pours, which would normally be wasted. The coping stones were produced over time meaning that no extra concrete was ordered for their production, saving around 1.2 tonnes of CO_2 and around £2-3,000.

Additional benefits. The big win from the project was the use of ggbs as a replacement in the concrete. This also has many other benefits, apart from the saving in CO_2, including lighter colour (desirable in this case), lower early thermal cracking, higher strength development over time and increased durability and increased workability. A potential disadvantage was noted as slower, early strength development/longer striking time, which can increase construction programme, hence only 50% replacement being used in the base, but in this case the concrete typically achieved a strength of 30 N/mm^2 at seven days.

These reductions are from only half of the project, the second phase, which was due to start later, is estimated to save at least an equivalent amount resulting in a final saving of around 100 tonnes, which is almost 20% of the estimated project footprint, as well as potentially £25,000 in costs. For future projects, the team is looking at saving further CO_2 by potentially reducing the criteria for crack control steel reinforcement.

(Reproduced with permission from the Environment Agency – Case study IEM01/2010/002.)

Appendix B – Energy checklist

Company name		Project title	
Location		Contract no.	

Energy			
Pre-construction and construction phase	**Yes**	**No**	**N/A**
1. Have energy objectives and targets been set for the project?			
2. Has an energy monitoring process been implemented to monitor and report energy consumption and carbon emissions, including business travel and transport?			
3. Have discussions with energy supplier(s) been made to ensure electrical supplies are connected at the earliest opportunity to avoid power from generators?			
4. Have site offices been fitted with practical, energy-saving devices, including low energy lighting (LED), passive infrared sensors for lighting, timer switches and thermostats for heating, hot water, and so on?			
5. Does the site induction cover the main energy efficiency issues (such as maintaining plant and switching off when not in use)?			
6. Are there signs in place advising site personnel to switch equipment off when not required?			
7. Is site office equipment set up to print efficiently (such as double-sided) and to a central location rather than to individual printers?			
8. Has the site considered renewable energy to contribute to powering the site accommodation?			
9. Is there a logistics/transport plan that considers efficient transport arrangements (such as the use of consolidation centres or just-in-time delivery)?			
10. Has the site considered green travel arrangements for reducing staff travel to site (such as car share and cycle schemes)?			
11. Are delivery vehicles switched off when being loaded and unloaded (unless needed to operate a Hiab or similar)?			
12. Would it be possible to use eco-friendly site cabins?			
13. Have the designers specified using low carbon concrete?			
14. Would it be possible to source materials from local suppliers?			
15. Are staff provided with electric or dual fuel vehicles?			
16. If electric cars are used on site are charging points available?			
Quality control and sign off	**Yes**	**No**	**N/A**
17. Have quality checks on the installed fabric insulation been completed?			
18. Did the building meet its target airtightness level?			
19. Have all renewable energy systems (such as biomass or PV) been fully commissioned and tested and is documentation in place to ensure that feed in tariffs (FITs) or renewable heat incentive (RHI) tariffs can be obtained?			
20. Have all building services, meters and control systems been fully commissioned and tested?			
21. Is a plan in place for monitoring during occupancy and seasonal commissioning?			
Comments			

Name		Position		Signature		Date	

For an interactive PDF version of this form visit the companion website.

E 04

Appendix C – Transport checklist

E 04

Company name		Project title	
Location		Contract no.	

Transport

	Yes	No	N/A
1. Have local, sensitive areas been identified (such as schools and residents)?			
2. Has the location of suitable parking arrangements for private cars and plant been defined?			
3. Has permission been obtained from the Local Authority for any road closures or erection of hoarding on the public highway?			
4. Have local pedestrian diversion routes been agreed with the Local Authority?			
5. Have delivery routes for construction traffic been agreed with the Local Authority?			
6. Have suppliers been made aware of any delivery restrictions and routes?			
7. Are entrance and exit gates on main roads, rather than side roads?			
8. Is there a process in place to keep roads clear of dust and dirt?			
9. Are deliveries scheduled to avoid traffic disruption or queuing outside of the site?			
10. Have designated vehicle routes on site been defined?			
11. Are vehicle routes designed to avoid excessive use of reversing sirens?			
12. Have deliveries been organised to avoid times of heavy traffic (for example, school pick up or drop off)?			
13. Is there a logistics/transport plan that considers efficient transport arrangements (such as the use of consolidation centres or just-in-time delivery)?			
14. Are communication processes in place to keep local residents informed?			

Comments

Name		Position		Signature		Date	

For an interactive PDF version of this form visit the companion website.

05

Water management

Contents

Supporting information

G: Checklists and forms

GT 700 Toolbox talks

Overview

Water is one of our most valuable natural resources, vital for our social and economic wellbeing and to maintain precious habitats. Despite the constant renewal of water resources, its supply is not endless. Therefore, we have a duty to ensure that it is protected and managed effectively.

This chapter gives a general overview of the legal framework for the protection of natural water resources, including the permit and licence requirements needed for making discharges to foul and surface water drainage systems.

This chapter also provides practical advice for the prevention of pollution to surface and groundwater and implementing an incident response plan should an accidental spillage occur as a result of construction work.

5.1 Introduction

Construction work cannot only cause serious harm to watercourses, plants and wildlife, it can also affect the quality and availability of drinking water resources and can be visually intrusive. Pollution can contaminate drinking water, suffocate fish and remove essential oxygen from the water, killing plants, animals and insects living in the water. A construction site does not need to be next to a watercourse to cause a problem; any pollutants entering a surface water drain can end up in a watercourse miles away.

Pollution by silt can result in the suffocation of fish, destruction of spawning sites and the blocking of drains, which in turn can lead to flooding. Silt pollution can be caused by dewatering, over-pumping, rainwater run-off from uncovered stockpiles of material exposed during earthworks, tunnelling operations, cleaning of ditches and drains and processes such as wheel washing.

Oil pollution reduces the levels of oxygen in water and can be toxic to aquatic wildlife. It coats plants, animals and birds. Oil pollution is mainly caused through spillages, often from refuelling, but can also be caused by accidental spills, vandalism and the overfilling of equipment. Oil, for example, spreads rapidly: one gallon of oil can completely cover an area of water the size of two football pitches.

Cement and concrete are probably the most common materials used in construction. If cement or concrete is allowed to enter a watercourse it can have a devastating impact on wildlife. Cement is highly alkaline and can alter the pH of the water, which can be toxic to aquatic wildlife and contaminate water supplies. Cement and concrete pollution is mainly caused by the cleaning out of equipment and shuttering failure.

Chemical pollution can have a wide range of impacts, including killing fish. Chemical pollution can be caused by spillages, the leaking of containers or incorrectly bunded areas.

Sewage pollution can be unpleasant, unsightly, smell and decrease the amount of oxygen in the water. Sewage pollution often occurs when drains are wrongly connected, blocked or damaged.

The **Environmental Protection Act** makes reference to controlling the entry of polluting matter and effluents into any place that may ultimately affect a watercourse.

The **Water Resources Act** makes it an offence to knowingly permit the pollution of controlled waters (such as those listed below) or contaminate water in a way that may poison or injure fish, spawn, fish food or spawning grounds.

- ☑ Rivers, streams, ditches, ponds, swales, underground streams, canals, lakes and reservoirs.
- ☑ Groundwater, wells, aquifers, boreholes or water in underground strata.

It is also an offence to deliberately or accidentally discharge trade effluents into public sewers without the relevant consent. Trade effluents are any liquids produced as part of a trade or industrial activity, excluding domestic sewage. Trade effluents include the water or slurry from vehicle wheel washers, core drilling, brick/concrete/stone cutting machines, dewatering trenches, pumping out of excavations, concrete washout, pipework cleaning and commissioning and any similar work.

Monitoring is an essential component to ensure watercourses are protected. Inspections confirm that the necessary preventative measures are in place and working efficiently. Monitoring must be undertaken on a regular basis to take account of changing site conditions or work methods. In periods of heavy rain, monitoring should be more frequent. Monitoring the effectiveness of the mitigation methods will inform of any required changes.

Just half a teaspoon of soil in a bath full of water would be comparable to water quality that could kill fish and smother plants in a watercourse.

5.2 Important points

- ☑ Before any work starts on site it is essential to identify all existing site drainage and provisions that have been made for pollution prevention for protection against oils, fuels and silts. Clearly mark these on site plans and distinguish which are surface water, foul water and combined systems (foul and surface water in the same sewer).
- ☑ Seek to install permanent drainage systems as early as possible, as these can then be used to avoid temporary discharges to surface water.
- ☑ All drains should be covered or protected to prevent accidental ingress from mud and silt.

- ☑ All stockpiled materials should be stored away from drainage systems and watercourses and protected using geotextile silt fencing or cut-off ditches where appropriate.
- ☑ The abstraction of water from surface water or piped mains (using a standpipe) will require consent from the Environment Agency (EA) for England, Natural Resources Wales (NRW) for Wales, the Northern Ireland Environment Agency (NIEA) for Northern Ireland, the Scottish Environment Protection Agency (SEPA) for Scotland, or the relevant water authority respectively.
- ☑ All discharges to foul water drainage systems require discharge consent from the local water authority and the consent conditions must be strictly complied with.
- ☑ All discharges to surface water systems will require an environmental permit from the relevant environment agency and the permit conditions must be strictly complied with.
- ☑ In England and Wales, temporary (less than three months) discharges from dewatering excavations can be carried out without the need for an environmental permit but strict conditions have to be complied with.
- ☑ Where septic tanks are designed to discharge to ground these will require an environmental permit for groundwater work. Discharges from small domestic septic tanks are exempt but this exemption is required to be registered with the EA or NRW.
- ☑ Before any discharges are made to surface water systems, the water must be unpolluted and free from silt. Silts can be removed through a variety of techniques, as listed below.
  - Settlement tanks.
  - Lagoons.
  - Filtration systems, including the use of gravels, geotextiles or straw bales.
  - Use of flocculants.
- ☑ All fuels and chemicals should be stored on impervious material away from drains and watercourses. They should be suitably bunded to prevent pollution in the event of leakage or spillage. Refuelling should also be carried out at designated locations away from drains or watercourses.
- ☑ All water from vehicle and boot washing facilities should be removed to foul water drainage systems (with the consent of the water authority) or taken away by tanker (waste duty of care must be complied with).
- ☑ Concrete and cement washout should not be allowed to enter surface water systems and should be carried out in designated areas. *(Further details on the disposal options for concrete washwater and the Environmental Permitting (England and Wales) Regulations are covered later in this chapter.)*
- ☑ Monitoring of all discharges should be made on a regular basis to ensure that consent conditions (quality and quantity) are being complied with. Oil and chemical storage facilities should also be inspected.
- ☑ Site and public roads should be regularly swept to reduce silt and mud entering surface water drainage systems.
- ☑ An incident response plan should be implemented identifying the following.
  - Type and location of drainage systems.
  - Type and location of spill kits.
  - Responsibilities for site personnel.
  - Awareness of environmental issues via training and induction.
  - Arrangements for spill kit replenishment.
  - Arrangements for disposal for contaminated materials from used spill kits.

5.3 Water scarcity

Water scarcity is both a natural and a human-made problem. There is enough fresh water on the planet for six billion people but it is distributed unevenly and too much of it is wasted, polluted and unsustainably managed.

Water scarcity is among the main problems to be faced by many societies in the world. Water use has been growing at more than twice the rate of population increase in the last century, and, although there is no global water scarcity as such, an increasing number of regions are chronically short of water.

Water scarcity is defined as the point at which the total impact of all users affects the supply or quality of water under existing arrangements to the extent that the demand by all sectors, including the environment, cannot be fully met. It can occur at any level of supply or demand and may be caused by human behaviour or the consequence of altered supply patterns (for example, from climate change).

Scientists typically measure scarcity by looking at the total amount of water available per person. An area is experiencing water stress when annual water supplies drop below 1,700 m^3 per person. When annual water supplies drop below 1,000 m^3 per person, the population faces water scarcity, and below 500 m^3 is absolute scarcity.

Water scarcity already affects every continent. Around 1.2 billion people, or almost one-fifth of the world's population, live in areas of physical scarcity, and 500 million people are approaching this situation. Another 1.6 billion people, or almost one quarter of the world's population, face economic water shortage (where countries lack the necessary infrastructure to take water from rivers and aquifers).

Whilst the UK as a whole does not suffer from water scarcity, areas in the South and East of England regularly suffer from serious levels of water stress. Levels of water stress have been calculated by the EA based on the following criteria.

- Current per capita demand for water.
- Current water resource availability.
- Forecast growth in per capita demand for water.
- Forecast resource availability.
- Forecast population growth.

(For further information on water efficiency and minimisation refer to 5.10.)

5.4 Groundwater

Groundwater is the largest available reservoir of fresh water and accumulates in gaps in the rocks called aquifers. Groundwater provides a third of our drinking water in England and Wales and also maintains the flow in many of our rivers. In some areas of southern England, groundwater supplies up to 80% of drinking water through the taps.

The Water Resources Act gives the EA or NRW a duty to protect the quality of groundwater and to protect it as a valuable water resource. The Environmental Permitting (England and Wales) Regulations replaced the Groundwater Regulations and implemented the requirements of the EU Groundwater Directives. Similar legislation is in place in Scotland and Northern Ireland.

Directive 80/68/EEC, on the protection of groundwater against pollution caused by certain dangerous substances (List I substances were prohibited and List II were only authorised with conditions), provided a groundwater protection framework before the current Directive 2006/118/EC. It required prevention of the (direct or indirect) introduction of high priority pollutants into groundwater and to limit the introduction into groundwater of other pollutants, so as to avoid pollution of groundwater by these substances. Directive 80/68/EEC was repealed in 2013 and Annexes I and II of 2006/118/EC are under review.

As the regulator, the EA has established groundwater vulnerability maps and source protection zones (SPZs) for 2,000 groundwater sources (such as wells, boreholes and springs) used for public drinking water supply. There are four levels of SPZs (shown below). The shape and size of a zone depends upon conditions of the ground, how the groundwater is removed and other environmental factors.

- **Zone 1 (inner protection zones).** An area where any pollution can travel to a borehole within 50 days. Zone 1 protection zones provide a 50 m protection radius of a borehole.
- **Zone 2 (outer protection zones).** Areas where pollution takes 400 days to reach a borehole, or when 25% of the total catchment area is affected, whichever is the biggest. This zone has a minimum radius of 250 or 500 m around the source, depending on the size of the abstraction.
- **Zone 3 (total catchment).** The total area needed to support the removal of water from a borehole.
- **Zone 4 or zone of special interest** was previously defined for some sources. Zone 4 (SPZ4) usually represented a surface water catchment that drains into the aquifer feeding the groundwater supply (for example, catchment draining to a disappearing stream). In the future this zone will be incorporated into one of the other zones, SPZ1, 2 or 3 (whichever is appropriate in the particular case) or become a safeguard zone.

Groundwater can be polluted by a range of construction materials, including fuels, chemicals, solvents, paints and other liquids. Pollution of groundwater will lead to prosecution. A small quantity of a pollutant has the potential to affect large volumes of groundwater. Remediating large volumes of groundwater is expensive and can take a long time.

Run-off from rainfall has the potential to pick up contaminants as it moves across a construction site. Contaminants may include particulates as well as hydrocarbons and chemicals. Run-off can penetrate permeable surfaces and infiltrate into groundwater.

Every effort must be made to prevent groundwater pollution, including the following.

- Proper materials storage with bunds protecting liquids stores.
- Suitable spill kits and competent spill teams with regular drills.
- Storing materials on impermeable surfaces.
- Use of interceptor drains to catch run-off before it reaches permeable surfaces.

The unauthorised discharge of chemicals or sewage to groundwater without a permit is an offence and could lead to prosecution.

5.5 Extraction

Water is often taken from natural sources or water mains to reduce dust on haul roads or to reduce dust during cutting operations.

Section 24 of The Water Resources Act states that:

> **... no person shall abstract water from any source of supply or cause or permit any other person to abstract any water, except in pursuance of a licence granted by the authority.**

You should therefore not take water from groundwater, watercourses, lakes, streams or water mains without the permission of the relevant authority. The EA, NRW, NIEA and SEPA are responsible for licensing abstractions from groundwater and watercourses. The local water supply company will be responsible for licensing water that is taken from water mains.

In England, Northern Ireland and Wales abstractions of up to 20 m^3 a day can be taken from watercourses and other natural water resources without an abstraction licence from the EA, NRW or NIEA.

In Scotland, under the Water Environment and Water Services (Scotland) Act and the Water Environment (Controlled Activities) (Scotland) Regulations (as amended), abstractions of up to 10 m^3 a day can be taken without registration with SEPA, subject to meeting the relevant general binding rules.

Water volumes abstracted should be monitored on a daily basis to ensure that the conditions of any abstraction licence are complied with. This will also allow you to check whether there are any leaks in the system.

5.5.1 Dewatering excavations

If, as part of the works, you are dewatering or pumping water that has gathered in an excavation, it does not require an abstraction licence if the water is to be disposed of solely to prevent interference with building operations.

If, however, you intend to use water from a dewatering operation for dust suppression or pressure testing on site, you may require an abstraction licence if over 20 m^3.

Construction dewatering is the temporary lowering of groundwater levels by pumping from wells or sumps to provide stable conditions for excavations below the natural groundwater level. The water being removed may require the appropriate consent for discharge into either the foul water system or controlled waters. Dewatering points must be sited to avoid other pollutants or water of a different chemical composition from entering a body of groundwater.

The EA and NRW are currently consulting jointly on their intended approach to remove most exemptions from water abstraction licensing and bring these abstractions under licensing control.

E 05

5.6 Disposal

Drains on site should be clearly identifiable as either surface water drains (marked blue), foul water drains (marked red) and combined sewers (marked red 'C'). Surface water drains carry uncontaminated rainwater directly to a stream, river or soakaway. Foul water drains carry foul water directly to a sewage works for treatment before being discharged to a watercourse.

5.6.1 Disposal from dewatering excavations

Water may enter an excavation from either surface water inflows or inflows of groundwater. The inflows may already be polluted or they may pick up pollutants contained within the excavation. The amount of water that is pumped out from an excavation can be reduced by reducing inflows of both surface and groundwater. Edge drains connected to sumps can intercept surface water flows. Cut-off ditches and well dewatering will reduce groundwater inflows.

Any pumping out into a trade effluent system (such as a sewer) will require consent from the water company. An environmental permit from the EA, NRW or NIEA will be required for a discharge to controlled waters (such as rivers, streams or lakes). However, in England and Wales you do not need an environmental permit for temporary water discharge if the discharge is for less than three months and if you can meet the requirements of the regulator's position statement for *Temporary water discharges from excavations*.

If you need a permit to discharge to a watercourse, it can take up to four months to obtain from the EA or NRW. *(For further details of consent requirements refer to 5.7.)*

Water being pumped to two blue silt traps (top left) allowing silt to settle prior to disposal

Treatment of water before discharge will reduce the potential impact that it has. Some forms of treatment are shown below.

- ☑ Pumping to grassland or other soakaway well away from excavations to avoid recirculation (this option is only suitable for unpolluted water containing only silt).
- ☑ Pumping to a settlement tank or lagoon, maximising retention time.
- ☑ Using a sump at the base of an excavation, wrapping the end of the pump in aggregates and keeping it off the excavation floor.
- ☑ Passing through a filtration system (such as aggregates, geotextile or straw bales).
- ☑ Using flocculants (chemicals used to aid solids removal) in conjunction with a settlement tank. Consultation with the regulators should take place first as adding chemicals could make things worse.

Where there is no alternative, water may also be taken and treated off site as waste and will therefore need to be controlled in accordance with the waste duty of care, together with the completion of waste transfer documentation.

5.6.2 Concrete and cement washout

Concrete and cement washout is highly alkaline and can cause severe pollution. Effluent produced from washing out any concrete mixing plant or ready-mix concrete lorries must not be allowed to flow into any drain, watercourse or to ground (groundwater). Washout areas should be designated and at least 10 m away from a watercourse and any drains.

A lined skip can be used to place waste concrete and effluent, with the water being pumped to a foul sewer or taken away by tanker. The concrete can also be recycled for reuse into the works.

A clearly designated washout area

Concrete wash water needs to be controlled

E
05

5.6.3 Disposal of sewage

The provision of welfare facilities at fixed and transitional construction sites requires that disposal of sewage must be considered. Where possible, disposal may be made by direct connection to a foul sewer. Direct connection to the foul sewer will require consent from the maintaining authority. On greenfield sites or sites remote from live foul sewers there may be no opportunity to connect to a local foul sewer, in which case a septic tank will be provided that can be regularly pumped out by a liquid waste disposal company. Where sewage waste is taken from site then the waste duty of care must be complied with, and waste transfer documentation completed.

Both the location and design of the septic tank are important considerations to avoid raw sewage from entering groundwater or watercourses. The use of portable toilet facilities should be discouraged wherever possible. Where effluent from septic tanks is designed to be discharged to ground then an environmental permit will be required.

5.6.4 Disposal from vehicle and boot washing

Where wheel-wash facilities are provided on site, the resultant water will be contaminated with silt and possibly oil from vehicle bodies. This effluent must not be discharged to surface water systems and should be removed to foul sewers with the consent of the local water authority. Modern wheel-wash facilities will allow the wash water to be recycled and recovery of the silts to be separately removed as waste. Where the water is contaminated it may be removed by tanker.

Vehicle wheels being inspected and washed before leaving site

Site boot cleaning facilities

Facilities should also be provided to allow site personnel to clean their boots before leaving site or entering site accommodation. The silty water from these facilities should be dealt with in the same way as the vehicle washing effluent.

5.6.5 Sustainable urban drainage system

Sustainable urban drainage systems (SUDs) can be used to manage surface water run-off from large areas (such as part of a housing estate, major roads or business parks). They provide a natural approach to managing drainage in and around developments. SUDs work by slowing down and holding back the run-off from a site, allowing natural processes to break down pollution. They deal with run-off close to the source rather than transporting it elsewhere.

They are designed to attenuate (storing and slowly releasing) surface water from developments in a manner that will provide a more sustainable approach than the previous, conventional practice of routing run-off through a pipe into a watercourse. They are also a tool for preventing flooding.

Facilities for SUDs include the following.

- ☑ Permeable surfaces.
- ☑ Filter strips.
- ☑ Filter and infiltration trenches.
- ☑ Swales.
- ☑ Detention basins.
- ☑ Underground storage.
- ☑ Wetlands.
- ☑ Ponds.

Other facilities exist (such as hydraulic controls or silt traps).

In England, Northern Ireland and Wales you may have to include plans for SUDs when you apply for planning permission for a development. It is good practice to include the use of SUDs in all development plans.

In Scotland the rules are considerably different and all new developments must use SUDs to control water run-off to the water environment, unless the run-off is from a single dwelling.

You must have a licence from SEPA if you plan to use SUDs for any of the following.

- ☑ A development with more than 1,000 houses or more than 1,000 car parking spaces.
- ☑ An industrial estate.
- ☑ Major roads and motorways.

For other developments you may not need to contact SEPA but you must comply with the requirements of the general binding rules (GBRs) 10 and 11 of the Controlled Activities Regulations (CAR).

5.7 Consents

5.7.1 Discharge consents

Where there is a requirement to discharge effluent from any construction activity to drainage systems, watercourses or rivers and streams, an application for consent to discharge must be made to the relevant authority. The issuing authority will depend on where the discharge is made. For example, discharges to foul sewers are usually regulated by the local water company under the provisions of the Water Industry Act. Here, a discharge consent from the appropriate water company is required, except where all discharges are regulated under an environmental permit (England and Wales) or integration pollution prevention and control (IPPC) authorisation (Scotland). Any discharges to surface water systems, rivers, lakes or ponds would be regulated by the EA, NRW, NIEA or SEPA.

Make sure contaminated water is contained

In England and Wales, consents to discharge to surface water systems are regulated under Schedule 21 of the Environmental Permitting (England and Wales) Regulations. The water discharge work covered by these regulations includes the discharge or entry to surface waters that are controlled waters (but not to groundwater) of any poisonous, noxious or polluting matter, waste matter, trade effluent or sewage effluent. The term *water discharge activities* also includes any work that results in deposits that can be carried away in water (such as cleaning the bottom of a river channel).

As highlighted earlier, temporary discharge from dewatering excavations does not require an application for an environmental permit provided that the:

- ☑ discharge is temporary, for an overall period of less than three months
- ☑ discharge is made to a surface water (such as a river, stream or the sea)
- ☑ discharge does not pollute the surface water or adversely affect aquatic life
- ☑ discharge location is not within, or less than 500 m upstream of, a river or European marine site or site of special scientific interest (SSSI), or within a site designated for nature conservation (such as national nature reserves (NNR), local nature reserves (LNR) and local wildlife sites)
- ☑ discharge does not cause flooding from the surface water
- ☑ discharge does not cause erosion of the banks or bed of the surface water.

Discharges of uncontaminated surface waters are not classed as a water discharge activity. However, you should discuss any proposed discharge of surface water with the regulator before any work takes place.

Water discharge activities that meet certain conditions are exempt from requiring a permit. Some of these are shown below.

- ☑ A discharge from a small sewage treatment plant discharging 5 m^3 or less of effluent per day and subject to meeting other stringent requirements.
- ☑ A discharge from a septic tank discharging 2 m^3 or less of effluent per day.
- ☑ Vegetation management activities.

Low risk sewage discharges that meet the relevant conditions can be regulated under an exemption or standard permit. However, all other applications for a discharge permit will be regulated under a bespoke permit.

An application for a **discharge permit** must include, but not be limited to the following.

- The place at which the discharge will take place.
- The nature and composition of the material to be discharged.
- The maximum amount of material that is likely to be discharged in any one day.
- The time period over which the discharge will take place.
- Details of any monitoring and testing arrangements.

In Scotland, discharge consents to surface and groundwaters are regulated through the Water Environment (Controlled Activities) (Scotland) Regulations (as amended) (CAR) which came into force on 1 April 2006.

CAR introduces three levels of authorisations proportionate to the type of risk for the activity.

- General binding rules.
- Registrations.
- Licences.

General binding rules (GBRs), set out in Schedule 3 of the CAR, represent the lowest level of control and cover specific low risk work.

E 05

Work complying with the rules does not require an application to be made to SEPA, as compliance with a GBR is considered to be authorisation. Since the operator is not required to contact SEPA, there are no associated charges. SEPA has prepared a practical guide to implementing the CAR.

In Northern Ireland discharge consents are managed by the Northern Ireland Water Management Unit under the Water (Northern Ireland) Order.

5.7.2 Works consents

On 6 April 2012, when a further phase of the Flood and Water Management Act was implemented, responsibility for regulating work on ordinary watercourses in most areas of England and Wales transferred from the EA to lead local flood authorities. Lead local flood authorities are unitary authorities where they exist and county councils elsewhere.

In England and Wales a flood defence consent is required from the lead local flood authority before building a flow control structure (such as a culvert or weir) on an ordinary watercourse. For consent to carry out any works within 10 m of a watercourse, an application must include plans, sections and details including any environmental mitigation measures.

You will require an environmental permit or exemption for flood risk activities from the EA or NRW to carry out any work in, under, over or adjacent (within 10 m) to a statutory main river.

In Northern Ireland works consent is required from the Rivers Agency, which is part of the Department of Agriculture and Rural Development.

In Scotland, as with discharge consents, if you carry out building and engineering work that significantly affects the water environment, this is regulated through the Water Environment (Controlled Activities) (Scotland) Regulations (as amended) (CAR) and you must either:

- comply with certain GBRs that apply to low risk work
- register your work with SEPA
- get a licence from SEPA.

5.7.3 Septic tanks

The discharge of any sewage from a septic tank to the ground will require an environmental permit for groundwater work unless it is a small domestic discharge, then it may be exempt and should be registered under an exemption. Septic tanks with discharges of less than 2 m^3 per day are exempt from environmental permitting as highlighted above.

5.8 Pollution prevention

A number of measures may be implemented to prevent spillages and reduce the risk of a pollution incident. The EA has a dedicated area on its website to assist in the identification and management of issues of pollution risk in construction and other industry areas.

For access to EA pollution prevention guidance for businesses visit the Government website.

For a water management and pollution prevention checklist refer to Appendix B.

5.8.1 Site establishment

When planning the site set-up the following should be considered.

- ☑ If the site is in a sensitive area (for example, near to a watercourse or in a SSSI). If so, restrictions are likely to be placed on the site (such as limited fuel/oil storage).
- ☑ Potential drainage on site (such as land drains, foul sewers, surface water drains and soakaways).
- ☑ The location of plant away from drains and watercourses, especially fuel storage, top soil storage and waste disposal areas.
- ☑ That haul roads must be at least 10 m away from a watercourse. Consider construction of gullies or ditches alongside haul roads and around the perimeter of the working area to collect and channel surface water.
- ☑ Whether the environmental regulator (EA, NRW, NIEA or SEPA) or relevant body has given permission to any consents or licences required.
- ☑ The placing of stockpiles and spoil heaps, which must be away from drains and watercourses (use geotextile silt fencing or cut-off ditches to avoid silt run-off where appropriate).

Cut-off ditch

When planning or undertaking construction work, consideration needs to be given to the history of the site and the surrounding areas.

The site on which construction work is to be undertaken may be in the path of ground contamination seepage from an adjacent factory, chemical store, buried waste or other process that may result in pollution.

Records of water pollution may exist, which will give an indication of possible health problems for workers. Obtain all available historical records, as any subsequent pollution may be attributed to the construction company and not to the originators of the pollution. You should also check if there is any sewage discharge upstream as this may cause issues if you are pumping that water.

E 05

5.8.2 Silt management

Silt pollution is easily identified by discolouration or cloudy water. Good practices to avoid problems include the following.

Silt trap using straw bales and geotextile

- ☑ Only stripping the minimum amount of land required.
- ☑ Use of cut-off ditches and geotechnical silt fences.
- ☑ Diverting clean water away from bare ground.
- ☑ Not pumping silty water directly into a watercourse.
- ☑ Diverting silty water away from drains and watercourses using sand bags, for example.
- ☑ Planning for the treatment of silty water when pumping out excavations or managing surface water run-off.

Silt can be removed by using settlement tanks, ponds or lagoons, by allowing silty water to infiltrate through large areas of grassy ground, geotextiles filters or straw bales. *(For further information covering disposal refer to 5.6.)*

5.8.3 Oil and fuel storage

The storage of potentially polluting materials and the refuelling of mobile plant near watercourses or bodies of water (within 30 m) should be prohibited as far as practicable.

Examples of safe storage for oil

The storage of all potentially polluting material should be within an impervious bund with a capacity greater than 110% of the total potential stored contents (for multi-tank bunds, the capacity must be 110% of the capacity of the largest tank or 25% of the total tank capacity, whichever is the greater). The maximum holding capacity should be painted on the side. All level gauges, filling valves and vents and filling nozzles (when not in use) should remain within the bund. All valves should be kept locked when not in use and made available to authorised and competent persons only.

For further information about oil storage facilities visit the GOV.UK website.

The transportation of fuel across the site in drums or other containers should be avoided as far as possible. All mobile plant, including but not limited to cranes, compressors, generators, tanks, and so on, should be maintained and operated such that all leaks and spills of oil are minimised. Oil storage facilities should be regularly inspected for integrity.

Consideration should be given to the actions to take in the event of a spill, should one occur, and the necessary arrangements that should be in place. A suitable refuelling area and training will be required.

Further information on oil storage and pollution prevention regulations can be accessed online.

5.8.4 Chemical storage

The floor area used for storing or decanting chemicals must not be permeable. Old or corroded drums will cause more problems than those in good condition. The following measures can be taken to minimise the risk of contamination.

Good chemical storage container

- Purchase chemicals in the appropriate-sized containers to avoid the need for decanting.
- Where decanting is necessary, have safe procedures that avoid any spillage.
- Provide relevant information, instruction, training and supervision to employees.
- The proper disposal of all products.
- Provide clear procedures and training for operatives to deal with accidental spillages.
- Make drip-trays available for plant that is known to be leaking environmentally damaging fluids.
- Have set procedures for the refuelling or replenishing of plant so that any spillage cannot permeate into the ground.
- Install bunding around all storage areas, even temporary fuel stores on construction sites.
- Maintain equipment or storage vessels in good condition.
- Get into the habit of only storing or using products that are needed, and only store such products in areas with impermeable floors without drain gullies.
- Maintain good housekeeping procedures and avoid the accumulation of litter or rubbish.

An emergency and incident response plan, appropriate to the size of the site and chemicals being used, should be in place in case of any spillages or pollution alerts *(refer to 5.9 for further details)*.

5.8.5 Mud

Mud from construction work has the potential to damage the environment. In wet weather it can enter surface watercourses and drains. In dry weather it can dry out and, as dust, become airborne with the potential to be carried some distance.

Muddy water should not be discharged to surface water drains

Mud can be controlled using road sweepers and by dampening down during dry weather.

5.9 Pollution incident response plans

Incident readiness and response begins with considering what emergency scenarios and incidents (source) may occur on a construction site, on what pathways the pollution can travel and what in the surrounding environment may be impacted by them (receptor).

Types of incident include fires, spills and leaks. For each one of these types of incident several different parts of the environment may be affected. For example, a fire will release products of combustion to atmosphere and water used for fire-fighting may enter the surface watercourse, trade effluent sewers and also percolate into the ground.

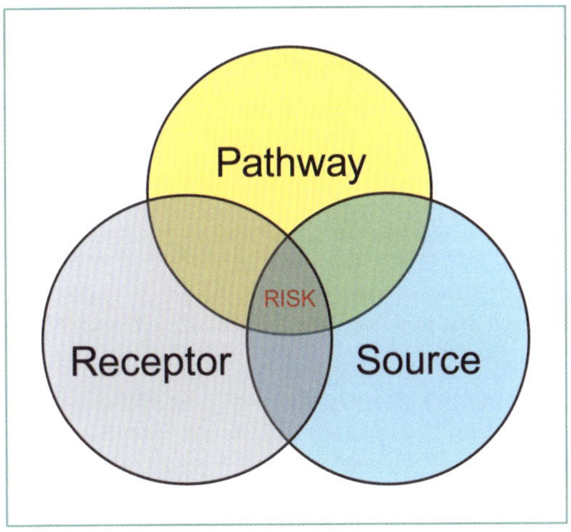

An effective incident response plan should include the following.

- ☑ Emergency contact numbers.
- ☑ Responsibilities of site personnel.
- ☑ Awareness of environmental issues via communication and training.
- ☑ The location of drainage systems or sewers and any arrangements for their stopping up if leakage into them occurs.
- ☑ The use of containment equipment (such as drip trays, bunds and booms) to avoid spillage.
- ☑ The location of spill kits around the site, including designated refuelling areas.
- ☑ Arrangements and responsibilities for spill kit replenishment.
- ☑ Arrangements for the safe disposal of used spill kit equipment and contaminated materials.

For each type of incident a plan should be put in place setting out the response needed. The basic process of **stop – contain – notify – clean up** must be followed.

In the event of a spillage it is essential that a competent spill response team is deployed with the right spill clean-up materials. The team has to respond sufficiently quickly for the spill to be controlled before it causes any environmental damage.

5.9.1 Booms

If work is to be carried out adjacent to a watercourse or river then arrangements should be put in place to ensure that a boom, which prevents the surface spread of oils and chemicals, can be deployed quickly to contain the spillage. The boom should be long enough to bridge across the river. If the work is carried out on navigable waterways then the appropriate authority should be notified before the boom is installed.

5.9.2 Drip trays

Where drip trays are used they should be deployed in conjunction with absorbent mats so that the contamination can be easily removed. Drip trays should not be left to fill with rainwater as this contaminated water will be more of a problem to dispose of and could cause contamination of drains and watercourses if left to overflow. Plant nappies are also available and allow water to pass through while absorbing any oils.

5.9.3 Spill kits

The type and quantity of spill kits deployed around a site will be dictated by the nature of the works and whether they are carried out on land or water. The type of spill kit will also depend on the type and quantity of oils and chemicals that are being used on the site. Typical spill-kit equipment could include, but not be limited to the items below.

Spill kit deployed

- ☑ Absorbent booms.
- ☑ Absorbent granules.
- ☑ Drain covers.
- ☑ Heavy-duty plastic bags.
- ☑ Absorbent pads.
- ☑ Shovel.
- ☑ Gloves.

There are three types of skill kits.

1. **Oil absorbents** that absorb oil and reject water.
2. **Chemical absorbents** that absorb water and soak and hold chemicals.
3. **Maintenance absorbents** that absorb water, oils and solvents

The correct type must be chosen to reflect the potential source.

Spill kits should be clearly marked and located at signposted locations around the site. Arrangements for the replenishment of the contents should be kept on the inside lid of the spill kit.

E 05

Equipment contaminated with oils and chemicals is likely to be hazardous waste and should be stored separately and disposed of in accordance with the duty of care and Hazardous Waste Regulations.

Spill response tests should be carried out on a regular basis that evaluates the effectiveness of incident response plans. According to the success of the response, improvements may be needed to the system. After any incident, management must conduct a root cause analysis to examine what improvements are needed to prevent the incident from happening again.

It is important to ensure that staff are trained in the use of spill kits.

For further information on pollution prevention for businesses visit the DEFRA website.

5.10 Water efficiency

Water is a precious global resource, critical for life in all its forms. As both the world's population and per capita water use increase there will be an ever-rising demand on what is a finite resource. Future climate change and resulting changing patterns of rainfall will make water supply increasingly challenging through the ageing water supply infrastructure.

Water is an expensive item with costs on both the supply side and the waste water treatment side. Given these pressures water will become a more expensive resource. What this means within construction is that water efficiency will become a higher priority at all stages in a building's life cycle. Materials with high embodied water content (such as high water usage during manufacture or use) will inevitably increase costs to take account of this priority. The increasing importance of embodied water is recognised in CEEQUAL, for example (physical resources use and management).

Whilst the abstraction of water from site excavations does not require an abstraction licence, abstracting water for other uses (such as dust suppression, washing down, and so on) will require a licence if using more than 20 m^3 per day in England and Wales.

There are a variety of techniques that can be used to increase water efficiency. The Strategic Forum for Construction (SFfC) has prepared an action plan, measurement protocol, water hierarchy and toolbox talk for managing and reducing water usage on construction sites. Measurement and understanding water performance is important so action can be taken to reduce consumption.

Refer to the SFfC website for its *Water: an action plan for reducing water usage on construction sites.*

For an example of a *water use in contruction* toolbox talk refer to Appendix A.

5.10.1 Using and conserving water during site works

On larger construction projects, where high volumes of water are being used, the first action is to establish an approach to measuring and monitoring water usage so that it can be managed. This may involve the use of water meters at appropriate locations and the use of water balances to account for water usage. Water reduction targets can be set based on known volumes of water usage and progress monitored.

Collecting data on water consumption during the construction process will provide the following benefits.

- ☑ Understanding and managing costs.
- ☑ Reducing environmental impact of overuse.
- ☑ Benchmarking and improving performance.
- ☑ Obtaining credits under BREEAM and CEEQUAL.
- ☑ Demonstrating continual improvement in accordance with ISO 14001/EMAS.
- ☑ Demonstrating good practice and meeting customer expectations.

Significant savings can be made by using rainwater harvesting systems to collect rainwater from roofs and other flat surfaces. Early installation of suitable collection systems would need to be investigated at the design stage and payback times calculated for the expected volume of water use. Harvested water can be used for dust suppression, avoiding the need to draw water from the mains or abstraction.

During supervisors' site inspections water use can be monitored and any obvious leaks and running hoses identified and dealt with. The use of triggers on hoses will prevent hoses from running whilst unattended.

Vehicle wheel-wash equipment is now available with water recycling and recirculation systems fitted. These will reduce the volume of water used and have the potential to save money. These systems work by providing a solids settling area combined with the use of flocculants to further precipitate solids out. The solids collected can be periodically removed.

Site accommodation can be fitted with waterless urinals, push taps and rainwater harvesting for toilet flushes.

Collecting rainwater for reuse

Waterless urinals

A roof area of 400 square metres will collect over 12,000 litres of water after 25 mm of rainfall.

Appendix A – Strategic Forum for Construction – Water toolbox talk example

STRATEGIC FORUM FOR CONSTRUCTION

Toolbox Talk: Water

Water use in construction

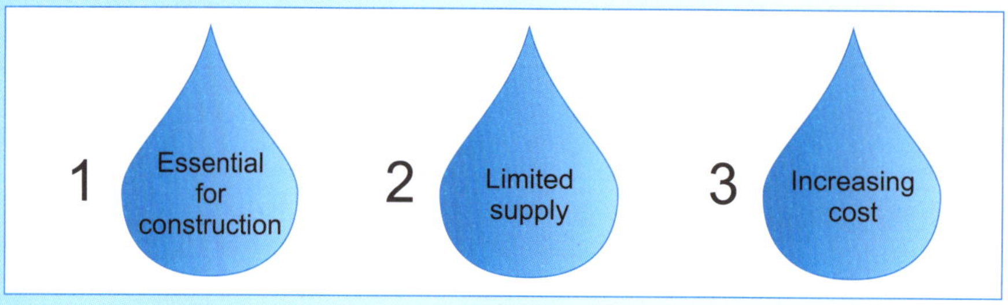

Water is integral to the economy, we need it for energy production, industrial processes, to grow food and, of course, for construction. In the coming years, the combined effects of climate change and a growing population are likely to put increasing pressure on our rivers, lakes and aquifers. If we do not act now to manage our demand for water, the security of our water supplies could be compromised.

What is the situation in the UK?

It is a misconception that the UK has plenty of water.
FACT - already, parts of England have less rainfall per person than many Mediterranean countries.
FACT - increasing demand will result in increasing cost both at home and on site as we fund new sources of supply.
FACT - water resources are under pressure and current levels of water abstraction are unsustainable in places.

What does this mean for construction?

- We can ensure no water is wasted.
- By reducing water usage, projects will benefit from cost savings.
- As an industry a commitment has been made to reduce water usage by 20% from a start position of 148m^3/£million contractors output. We all have a responsibility to measure progress against this target.
- We will be able to identify if water from other sources might be an appropriate alternative to using water of drinking quality standard.

STRATEGIC FORUM FOR CONSTRUCTION

Toolbox Talk: Water

E 05

Water Hierarchy:

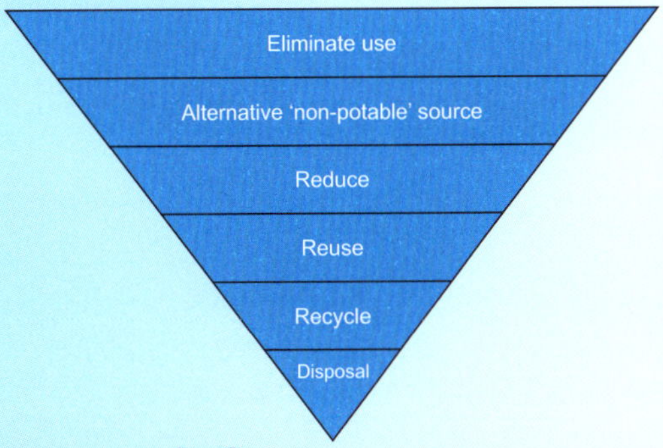

Is the process or activity really necessary? Is there a cost effective alternative to water?

Eliminate inappropriate use of drinking (potable) water. Can you use rainwater or grey water for the activity/process?

Explore options to improve efficiency. Can fittings or processes be updated?

Can water be treated/filtered for reuse in a process or activity?

Can water be recycled for use elsewhere?

Dispose of excess water legally and responsibly to ensure there is no flooding, pollution or inconvenience to others.

What can you do?

Hold a discussion with your team to identify where you use water on site. Refer to the 'SFfC How to Save Water on Construction Site' guide for the top ten quick reducing water use. Three of the most commonly applicable water saving measures are listed below;

1. Fix Leaks

£

An unfixed leak can be the most significant water use on site.
Leaks can come from damaged washers in taps, worn valves and corroded or damaged pipework.

2. Fit trigger guns to hoses

£

Hoses left running when not in use waste a lot of water in a short time.
Fit robust trigger guns to hoses so that flow can be controlled at the point of use.

3. Running taps

£

Flow from taps is often more than is needed. Consider adapting taps by either fitting a flow restrictor or aerating tap insert. Changing the tap is another option. Turn taps off.

Take away message

- Apply the water hierarchy. Prevent use, improve efficiency, consider alternative sources, reuse and recycle.
- Water is our most precious resource and every one of us has responsibility to conserve it.

Appendix B – Water management and pollution prevention checklist

Company name		Project title	
Location		Contract no.	

Water management and pollution prevention

	Yes	No	N/A
1. Have all watercourses and drainage systems been identified on site?			
2. Have all works in, above or near to watercourses been agreed with the relevant Flood Defence Authority (Local Authority or regulatory body, in the case of statutory main rivers)?			
3. Have all discharges to streams, ditches and drainage systems been consented to by the relevant environment agency?			
4. Where required, have all water abstractions from rivers, ponds, lakes or water mains been consented to by the relevant environment agency?			
5. Are site personnel inducted and suitably trained in dealing with waste water on site?			
6. Have designated areas been defined to wash out concrete lorries away from watercourses and drains?			
7. Have water discharges been properly treated (such as using settlement tanks or lagoons)?			
8. Have water monitoring procedures been put in place to ensure discharges are the correct quality?			
9. Are watercourses and drainage systems protected from run-off and silty water?			
10. Are all fuel, oil and chemical tanks effectively bunded to at least 110% of their capacity (or 25% of total capacity for drums)?			
11. Are all fuel, oil and chemical tanks located as far as possible from drains and watercourses?			
12. Are all fuel, oil and chemical tanks separated from the ground by an impermeable layer?			
13. Are all fuel, oil and chemical tanks marked with the type of contents, volume and appropriate hazard warning signs?			
14. Are steps being taken on site to prevent ground contamination or pollution by fuels, oils, chemicals, paint, and so on?			
15. Have site personnel been made aware of the site spillage response procedures through inductions, training and toolbox talks?			
16. Are appropriate spill kits available and appropriate personnel trained to deal with any accidental spillages to drains or watercourses?			
17. Have proactive measures been taken to reduce water consumption (such as grey water recycling) and water saving devices (such as waterless urinals and triggers on water hoses)?			
18. Have water meters been fitted to measure water usage?			
19. Is water consumption being monitored and recorded and communicated to site personnel to promote water minimisation?			

Comments

Name		Position		Signature		Date	

For an interactive PDF version of this form visit the companion website.

E 05

06

Statutory nuisance

Contents

Supporting information

G: Checklists and forms

GT 700 Toolbox talks

Overview

Construction sites are inherently noisy and can generate dust, exhaust emissions and fumes. Noise is the largest single source of complaint but consideration should also be given to the effects that lighting, vibration, traffic movement and odours can have on neighbours of construction sites.

This chapter identifies construction processes that could cause nuisance, provides an overview of the legal framework for their control and provides guidance on the mitigation and management of these issues on site.

6.1 Introduction

The quality and enjoyment of people's local environment have been recognised in law for many centuries. Construction sites, (which generate dust, noise, fumes and artificial light) have the potential to cause nuisance to neighbours. This must be managed effectively to avoid Local Authority or individual intervention, with the potential of delaying or stopping the works.

There are three types of nuisance: statutory nuisance, where a particular nuisance has been recognised in statute law, and public and private nuisance, which are within common law.

Current legislation for statutory nuisance in England, Scotland and Wales is contained in Part III of the Environmental Protection Act, and is enforced by Local Authorities with controls on various types of noise nuisance regulated under the Control of Pollution Act.

Section 79 of the Environmental Protection Act includes the following statutory nuisances.

- ☑ Premises in such a state as to be prejudicial to health.
- ☑ Smoke emissions that are prejudicial to health or a nuisance.
- ☑ Fumes or gases emitted from premises so as to be prejudicial to health or a nuisance.
- ☑ Any dust, steam, odour or other waste that is a nuisance or prejudicial to health.
- ☑ Any accumulation or deposit that is prejudicial to health or a nuisance.
- ☑ Artificial light emitted from premises so as to be prejudicial to health or a nuisance.
- ☑ Noise emitted from premises that may be prejudicial to health or a nuisance.
- ☑ Noise emitted from, or caused by, a vehicle, machinery or equipment on a street or road.
- ☑ Any water covering land, or land covered with water, that is in such a state as to be prejudicial to health or a nuisance (Scotland only).

Local Authorities have a duty to inspect their areas from time to time to detect whether a nuisance exists and investigate all complaints of statutory nuisances. Local Authority lane rental or permit schemes may push work into off-peak times during the night, thus increasing the possibility of creating a nuisance.

Where a Local Authority is satisfied that a statutory nuisance exists or is likely to occur then, under Section 80 of the Environmental Protection Act, the Local Authority can serve an abatement notice on the person responsible for the nuisance. If the person cannot be found then the notice is served on the owner or occupier of the premises.

Failure to comply with the terms of an abatement notice may result in prosecution in a Magistrates' Court. Conviction may result in a fine of up to £5,000 plus a daily fee of £500 for each day the offence continues after conviction. Under Section 82 of the Environmental Protection Act, individuals can also take action through the Magistrates' Court.

All employers should ensure that best practical means have been used to prevent, or to counteract the effects of, the nuisance.

Best practical means

Best practical means is defined as taking the following points into account.

- ☑ Current technical knowledge.
- ☑ The design, construction and maintenance of buildings and enclosures.
- ☑ Design, installation, maintenance and periods of operation of plant.
- ☑ Financial implications.
- ☑ Local conditions.

6.2 Important points

Effective planning of site activities should take place to ensure that dust, fumes, odours and noise do not cause annoyance to neighbours or impact on wildlife.

To avoid causing complaints sites should adopt the following practices.

- ☑ Identify sensitive receptors (people, wildlife and the environment), liaise with Local Authorities on any nuisance that is likely to occur and agree the appropriate mitigation measures.
- ☑ All transport activities, both on and off site, should be planned to avoid dust, noise, fumes and disruption on local roads.
- ☑ Carry out regular damping down of haul routes and cleaning of the public highway.
- ☑ Locate any crushing or cutting equipment away from sensitive areas and dampen down, as appropriate.
- ☑ Store dusty materials under cover.
- ☑ Consider alternative methods of construction to reduce dust and noise.
- ☑ Where noise could be an issue, apply to the Local Authority for a Section 61 prior consent to agree construction methods and noise levels.
- ☑ Locate noisy equipment away from sensitive receptors and/or use screening, where necessary.
- ☑ Ensure that plant and equipment are regularly maintained and kept in good working order.
- ☑ Ensure that a monitoring and inspection regime is in place to quickly identify and rectify problems.
- ☑ Prepare and communicate the process and responsibilities for dealing with and responding to environmental complaints.

6.3 Airborne dust, emissions and odours

A variety of air pollutants have known or suspected harmful effects on human health and the environment and can create a statutory nuisance. The UK Government has made provisions to ensure that air quality standards for certain pollutants are set centrally through the national air quality strategy.

Generally, Local Authorities have control over the management of air quality through the identification of air quality management areas. In addition, Local Authorities regulate the smaller, less polluting, installations and processes (such as batching plants, crushing work or paint spraying activities). The Environment Agency (England), Natural Resources Wales (Wales), the Northern Ireland Environment Agency (Northern Ireland) and the Scottish Environment Protection Agency (Scotland), on the other hand, regulate air emissions from the more polluting processes (such as landfill sites and asphalt plants).

6.3.1 Controls under the Clean Air Act

The Clean Air Act provides a comprehensive control mechanism for the protection of the environment from smoke, dust, grit and fumes from all fires and furnaces, with the Local Authority as the relevant environmental regulator who can impose limits on dust, emissions and odours generated from a site.

Under Section 1 of Part 1 of the Clean Air Act, it is an offence to emit dark smoke from the chimney of any building on any day unless these occur within the permitted periods, as specified in the Dark Smoke (Permitted Periods) Regulations.

Under Section 2 of Part I of the Clean Air Act, it is an offence to cause, or permit the emission of, dark smoke from industrial or trade premises (other than chimneys); premises could also include a demolition site. The method for assessing the shade of smoke is based on the Ringlemann chart. This determines the type of smoke depending on its shade. Only when smoke colour reaches the 40% obscuration point is it considered dark.

6.3.2 Controls under Environmental Permitting Regulations or Pollution Prevention Control Act

The provisions of the Clean Air Act do not apply to processes that are controlled under the Environmental Permitting Regulations in England and Wales, the Pollution Prevention and Control (Industrial Emissions) Regulations in Northern Ireland or the Pollution Prevention Control Act in Scotland. From a construction perspective these include, but are not limited to, installations such as batching plants, crushing work, waste transfer stations and landfill sites.

The standard conditions for a Part B permit for mobile crushing plant requires that no dust must cross the site boundary and the Local Authority must be informed if it does. In addition, visual inspections of dust must be made three times daily.

It is also possible to obtain an environmental permitting D7 exemption from the Environment Agency or Natural Resources Wales for the burning of plant waste on land in the open (this only applies to the burning of waste by an establishment or undertaking where the waste burned is the establishment or undertaking's own waste). This exemption limits the burning of waste of up to 10 tonnes in 24 hours and could include, for example, Japanese knotweed.

6.3.3 Controls under Highway Regulations

The Highways Act, Sections 161 and 161A, forbids the lighting of a fire on or above a highway, within 50 feet of the centre of the highway, without authorisation. If a fire is lit on land that is not part of the highway but consists of a carriageway, or a person allows this to happen, and a user of the highway is injured, interrupted or endangered by the fire, smoke or any other fire resulting from the original fire, then those responsible for the fire will be liable to a fine unless reasonable attempts were made to prevent the fire from occurring.

Managing dust and emissions

Fine water mist being sprayed to reduce airborne dust clouds

- ☑ Identify sensitive receptors and liaise with the Local Authority regarding any likely nuisances that could occur.
- ☑ Put control measures in place to mitigate any negative dust impacts, including the following.
  - Dampening down of haul routes with water.
  - Ensuring public highways are regularly swept.
  - Installing wheel-washing equipment at site exits where appropriate.
  - Ensuring that bulk materials leaving site are covered.
  - Installing dust screens or silt fencing to prevent dust spreading.
  - Using mist cannons in dust suppression *(as shown in the image)* and water suppression in cutting and grinding work.
  - Ensuring that bulk materials likely to cause dust are covered where appropriate.

E 06

6.3.4 Silica dust

Silica is a constituent of sand, and many construction activities produce high concentrations of silica dust. Respirable silica dust is harmful to health and exposure should be prevented.

Work, including concrete drilling, scabbling, chasing, cutting and sand or grit blasting techniques, can create large volumes of dust.

Clouds of dust do not restrict themselves to the boundaries of the construction site. They often migrate and contaminate the environment around the site.

Such pollution may cause problems for people in local food processing companies, restaurants, cafés, schools, hospitals and general living accommodation, and can also contaminate local watercourses or drainage systems and affect wildlife and plants.

For further information on the health risks associated with silica dust refer to the dust and fumes chapter in Section B.

6.3.5 Measuring and monitoring dust

There are currently no UK limits for assessing deposited dust and its ability to cause a nuisance. Reference is made to an annual deposition of 200 mg/m^2/day as a value for the threshold for serious nuisance. In addition, DEFRA's *Process guidance note PG 3/16* for crushing work does not specify any limits but simply refers to the avoidance of visible emissions crossing the site boundary. Visual assessment of dust is therefore a main indicator as to whether dust may cause a statutory nuisance.

Dust deposit gauge

Care should be exercised when specifying this as a proposed limit because it:

- ☑ does not consider the nature of the dust
- ☑ is thought to have essentially been derived using the BS deposit gauge and is not equally acceptable to all types of deposit gauge
- ☑ is an annual average and is therefore not applicable to shorter measurement periods
- ☑ has no statutory effect.

Deposit gauges are a simple and accurate method of measuring deposited dust. A dry frisbee dust deposit gauge can be used to determine the amount of dust produced a month at a time. It uses a bowl and bottle to collect large and small dust particles and should be situated 5 m away from any obstruction.

For further guidance from DEFRA on air pollution and control visit the Government website.

6.3.6 Exhaust emissions

Because of the fuel they burn, motor vehicles or other internal combustion engines, generators and compressors are among the largest sources of airborne pollution. This pollution is increasing steadily as the use of vehicles increases.

Elements from exhaust emissions that pollute the environment include the following.

- ☑ Carbon dioxide.
- ☑ Carbon monoxide.
- ☑ Hydrocarbons.
- ☑ Lead.
- ☑ Nitrogen oxides.
- ☑ Particulate matter (smoke).

The following actions can be taken to minimise the extent of pollution.

- ☑ Limit the use of road vehicles or other internal combustion engines.
- ☑ Ensure that vehicles are switched off when not required.
- ☑ Ensure that haul routes are planned with minimum distances.
- ☑ Have a planned and preventative maintenance programme or modify existing engines to produce less pollution.
- ☑ Use fuels designed to cause less pollution.
- ☑ Consider the use of electrical equipment rather than internal combustion engines.
- ☑ Liaise with electricity suppliers early in a project so that equipment can be connected to the grid as early as possible rather than running off generators.

6.3.7 Vapours and fumes

Many materials or products (such as sealants or resins) when used in a work environment may release vapours, fumes or odours. These can damage the environment and be hazardous to the health of workers or other persons.

The likelihood of such hazards must be assessed, and adequate control measures designed and implemented that include appropriate monitoring arrangements. The controls should eliminate the risks, where possible, or otherwise minimise them. The selection of less hazardous products should be considered as this may also avoid the used packaging needing to be disposed of as hazardous waste.

For a statutory nuisance checklist refer to Appendix A.

6.4 Noise, including consent requirements

Construction and demolition sites are inherently noisy and are often located in residential areas that are normally quiet. They have the potential to create a statutory nuisance in the form of noise and vibration that disturbs wildlife, causes structural damage to buildings and utilities and can create health risks to site staff and the general public.

Local Authorities may place restrictions on the person responsible for a construction site to observe specified controls designed to minimise noise and vibration nuisance.

The Control of Pollution Act, Section 60, gives Local Authorities the power to serve notices that specify the following.

- ☑ The maximum levels of noise that may be emitted from any particular point.
- ☑ Provisions for any change in circumstances.
- ☑ The type of plant or machinery that may, or may not, be used.
- ☑ The working hours when noise may be made.

BS 5228, which deals with noise and vibration control on construction sites, includes good practice to ensure that a Section 60 notice is avoided. It is in five parts, with Part 1 being a Code of Practice for basic information and procedures for noise and vibration control.

The Control of Pollution Act and BS 5228 do not specify any limits for construction noise on the basis that a Local Authority knows its area best and should have the best understanding of suitable noise limits.

6.4.1 Application for Section 61 noise consent (prior consent)

Where it is possible that a noise or vibration nuisance will be created, the person responsible for the site may make an application to the Local Authority for prior consent to start work. Such consent applications should, where possible, be made at the same time as planning applications, or as soon as it is known the nuisance cannot be avoided (such as power-floating at night). The Local Authority has 28 days to approve the application.

Applications should contain particulars of the following points.

- ☑ The work to be undertaken.
- ☑ Location of the works.
- ☑ Working hours.
- ☑ Proposed methods, and the plant and machinery to be used.
- ☑ Proposed steps for minimising noise and vibration.

The Local Authority, in granting consent for the works to begin, may:

- ☑ attach any conditions they wish to the consent
- ☑ limit or qualify a consent
- ☑ limit the duration of consent for the works to be carried out
- ☑ specify maximum boundary noise level, permitted hours of work, and plant and equipment that may or may not be used.

In the case of works that overrun for legitimate engineering or health and safety issues you should advise the Local Authority as soon as is reasonably practicable of the reasons for the over run, and the likely duration of such works.

Where there are minor variations in the works featured in the consent application, and rescheduling of works are of a critical nature, you may apply for a variation. This procedure may also be used for minor additional work that was not included in the original application and does not materially affect the predicted noise levels.

The Local Authority should receive applications for a variation, where practicable within seven days, but at least two working days ahead of the start of the works for which the application is made.

Where the proposed works have to be changed from the original programme, as given in the application to require operations outside of the terms of the consent, you must apply for a dispensation at least 14 days in advance of the proposed operation, submitting the following.

- ☑ Details of the operation in question.
- ☑ Reasons why the operation cannot be carried out within the terms of the consent.
- ☑ Proposed working hours.
- ☑ Predicted noise and vibration levels at relevant locations.
- ☑ Proposed steps taken to reduce noise and/or vibration to a minimum.

It is far better to apply for a Section 61 notice and work with the Local Authority to agree how the work will be carried out, than to be stopped following a complaint and have to comply with a Section 60 notice. Once a Section 61 consent has been given, the contractor is protected from action on noise grounds taken by the Local Authority as long as the conditions of the consent are complied with.

6.4.2 Managing noise on site

Before works commence on site the contract documentation should be reviewed to determine whether specific noise limits at various locations from the site boundary have been specified. Where this is the case, it is likely that monitoring of noise levels will be required and may also be a requirement of a Section 61 noise consent, where this has been applied for.

Periodic noise survey

A noise survey will establish the ambient and background noise levels at relevant locations around the site boundary. The purpose of the noise survey will be to establish the best location for noisy operations that could cause a nuisance to neighbours; the survey can also be used to provide supporting evidence for invalid claims.

The control of noise on construction sites can be achieved either by controlling the noise at source or by screening.

Methods for **controlling noise at source** are shown below.

- ☑ **Selection of low noise method.** Where possible, methods should be employed to reduce the amount of noise generated in the first place (for example, off-site fabrication of concrete panels would avoid the need for scabbling of concrete and the use of vibrating pokers).
- ☑ **Working hours.** Adapt working hours to restrict noisy activity to certain periods of the day. Arrange delivery times to suit the area.
- ☑ **Selection of quiet or low noise equipment.** For some types of noisy operations (such as piling) there are alternative methods available (for example, drop hammer piling could be replaced by hydraulic jack if ground conditions are suitable). Many power tools are now available that can be operated using electricity or compressed air rather than petrol or diesel engines.

- **Location of equipment on site.** Where possible, noisy stationary equipment should be placed away from sensitive receptors and public areas.
- **Provision of acoustic enclosures.** Most modern equipment (such as compressors or generators) will come with its own hood or door. These should always be kept closed and in good order. Acoustic enclosures can be purchased that surround equipment to reduce noise transmission.

Site fencing incorporating acoustic barrier to protect the public

Methods for **screening noise** include the use of the following items.

- Site hoarding.
- Purpose-built screens.
- Material storage.
- Bunding.
- Existing structures.

The screen should be placed close to either the source or the receptor.

Other general measures that can be employed to reduce noise levels are shown below.

- Planning site haul routes to avoid vehicles reversing.
- Planning delivery times and routes to suit local conditions.
- Maintaining haul routes in good order to prevent vehicle noise caused by potholes or uneven surfaces.
- Minimising drop heights of materials into lorries and dumpers.
- Shutting down plant when not required.
- Using only plant conforming with relevant standards and directives on emissions.
- Maintaining plant in good order, including compressor air lines.
- Placing material handling areas away from sensitive receptors.
- Making use of noise-reducing equipment (such as jackets, shrouds, hoods and doors) and ensuring that they remain closed when the equipment is in use.
- Ensuring all viewing openings in site hoardings are glazed with Perspex.
- Providing good practice guides to all operatives through the provision of toolbox talks.

6.5 Vibration

In simple terms, vibration is caused by sound waves travelling through solid material rather than through air. High levels of vibration can cause damage to buildings, disturb wildlife and disturb neighbours. In the UK, vibration is considered in the same manner as noise as regards statutory nuisance.

The most commonly used standard for environmental vibration assessment for disturbance is BS 6472 *Evaluation of human exposure to vibration in buildings [1 Hz to 80 Hz]*. The British Standard suggests that levels of vibration from 0.1 to 0.2 mm/s (at night) and below in residential buildings would have a very low probability of adverse comment.

BS 7385: Part 2 *Evaluation and measurement for vibration in buildings* gives guidance on the levels of vibration above which building structures could be damaged. The standard states that there is a major difference between the sensitivity of people in feeling vibration and the onset of levels of vibration that damage the structure.

For residential buildings, the standard states that, for cosmetic damage (such as cracking in plaster work) to occur, a peak particle velocity of some 15 mm/s is necessary at a vibration frequency of 4 Hz; this rises to 20 mm/sec at 15 Hz, and thereafter the limit rises to 50 mm/s at 40 Hz and above.

The potential for vibration is dependent on a number of factors, particularly the distance of the receptor from the source, together with the ground conditions and features within the ground (such as sewers). Piling, vibrating rollers, tunnelling and boring are important work that has the potential to cause vibration. Consideration should be given to the methods used to establish the solution with the least vibration risk. These types of activity should be considered carefully when working near to known archaeological features.

Hydraulic pressed sheet piling

Important questions to be considered when planning the works are listed below.

- ☑ Can the activity be done using a different technique that results in lower vibration levels (for example, hydraulic pressed sheet piling rather than impact piling or vibrating methods)?
- ☑ As high frequency vibration causes less damage than low frequency vibration, can the plant be operated in a way that generates less low frequency vibration?
- ☑ Can the equipment be isolated from the transfer medium (for example, putting generators on timber mats rather than directly on the ground)?

Before starting work, all sensitive structures and buildings should be identified and surveyed. Checks should be made to see if the work is covered under the Party Wall etc. Act. The survey should include photographic and written records of any of the following existing issues.

- ☑ Cracks and their width.
- ☑ Levels and verticality of tilting walls and bulges in walls.
- ☑ Damage, including broken bricks, tiles, pipework or plaster.

Where some vibration cannot be avoided you should adopt a good neighbour policy and inform local residents and the local environmental health officer.

Vibration levels should be monitored during the works using competent, trained staff with the appropriate equipment. Following the works, the same survey should be carried out to confirm that no damage has occurred.

6.6 Light pollution

Light pollution is artificial light that illuminates or intrudes upon areas not intended to be lit. There are a variety of sources of light pollution from construction sites, including lighting towers, offices and access or security lighting.

Check the direction of lighting and reflection off surfaces. Simple adjustments may easily reduce any light pollution affecting neighbours. Where possible lighting should be switched off to avoid disturbance.

In England, Northern Ireland, Scotland and Wales light pollution can be a statutory nuisance and Local Authorities have powers to apply an abatement notice. If this is not complied with the matter may then go to a Magistrates' Court. In Scotland the Public Health etc. (Scotland) Act introduced measures covering light nuisance that were implemented by amendment to the Environmental Protection Act. Similarly, light nuisance provisions were introduced in Northern Ireland by the Neighbourhoods and Environment Act (Northern Ireland).

6.7 Community liaison

A good neighbour has consideration for those around them and takes an active interest in their wellbeing. Construction sites may have many different types of neighbour (such as householders, businesses, sports clubs, pubs and shops) and construction work will affect neighbours in various ways.

Good relationships with the local community are vital

Construction projects need to take a proactive approach to community liaison and ensure that the right processes are in place for engaging with the local community. It is important to inform neighbours of what may affect them before it happens so that people can prepare and have a chance to have their say. There are a number of ways in which this can happen, including liaising with neighbours, holding meetings, school visits, site tours or open days, door to door visits and providing regular newsletters.

It is also important that an effective complaints system is in place that provides a rapid and effective response to issues that have been raised. Being a good neighbour will result in a good relationship with the local community and will avoid complaints and damage to reputation.

Started in 1997, the **Considerate Constructors Scheme** (CCS) was set up by the UK construction industry to improve its image. Since then, the scheme has registered and monitored over 70,000 sites and has been instrumental in many of the improvements enjoyed today.

The scheme is recognised by the UK construction industry as a major force in improving its image through the registration and monitoring of UK sites. The scheme is also recognised in, and contributes to, improved performance in certification schemes, including BREEAM, CEEQUAL and Ska.

Any work that could be considered by the general public as construction can be registered as a site, provided it has a duration of six weeks or more. A site registers onto the scheme and agrees to follow the code of considerate practice, which forms the basis of all of the scheme's requirements.

The site is then visited by a monitor, assessed and scored against the Code of Practice.

The monitor also identifies any measures being taken by a company or its sites that are above and beyond these requirements. Certificates of compliance and certificates of performance beyond compliance are awarded to sites when they meet the code scoring criteria. There are also national awards for the top performing sites and companies.

www.

For further information on the code of considerate practice, checklists, report formats and scoring system for both site and company registration visit the website.

Appendix A – Statutory nuisance checklist

Company name		Project title	
Location		Contract no.	

Statutory nuisance	Yes	No	N/A
1. Has the appropriate liaison or communication taken place with local stakeholders that may be affected by nuisance?			
2. Are appropriate dust suppression techniques used to minimise air pollution from timber sawing or planing, stone or block cutting, crushing, and so on?			
3. Is dust on site haul roads and material stockpiles dampened down adequately on dry, windy days?			
4. Are haul roads located away from sensitive areas (such as rivers and ditches)?			
5. Are site vehicle speed limits controlled to reduce dust?			
6. Are public roads regularly cleaned using a road sweeper or vacuum?			
7. Do vehicles that remove granular or dusty materials have sheeted covers?			
8. Are all plant and vehicles in good working order with an up to date maintenance or service log?			
9. Are enclosed chutes and covered skips used for lowering dusty demolition or waste materials?			
10. Are cement and concrete being mixed in enclosed areas to prevent dust?			
11. Are material stockpiles or spoil heaps protected and stored away from sensitive areas (such as drains, rivers and ditches)?			
12. Have the works been assessed to identify the noise and vibration impact on local neighbours, wildlife and buildings?			
13. Has the local environmental health officer and neighbours been consulted and forewarned of any out-of-hours or major disruptive activities?			
14. Have you applied for a Section 61 notice under the Control of Pollution Act 1974?			
15. If a Control of Pollution Act Section 61 consent is in place, are noise levels being recorded to ensure that levels are kept within limits?			
16. If possible, have working methods been reviewed to use equipment that reduces noise and vibration (such as pile jacking and chemical bursting)?			
17. Have working hours been defined to restrict noisy operations to certain times of day?			
18. Is noisy plant kept as far as possible from sensitive receptors?			
19. Are deliveries planned to suit the local area?			
20. Are haul routes well maintained to prevent vehicle noise and vibration?			
21. Where required, are noise screens being used to reduce noise transmission?			
22. Are noise screens and hoarding well maintained with no holes and gaps?			
23. Has lighting been positioned to avoid a nuisance at night and is non-essential lighting switched off at night?			
24. Does the site have a procedure for dealing with complaints and issues?			

Comments							
Name		Position		Signature		Date	

E 06

For an interactive PDF version of this form visit the companion website.

07

Ecology

Contents

Supporting information

G: Checklists and forms

GT 700 Toolbox talks

Overview

There is a high level of protection given to wildlife through legal controls and contract conditions.

This chapter gives a general overview of the legal framework for the protection of wildlife and their habitats. It highlights endangered and protected species and designated sites for their protection.

This chapter also gives an overview of the legal framework for the regulation of invasive species.

7.1 Introduction

The identification and management of wildlife needs to be undertaken early in the planning stage of a project to avoid costly delays to the programme and possible loss of reputation if damage takes place.

Damaging, disturbing or removing protected species can result in prosecution under a range of environmental legislation, and wildlife is also held in high regard by the public.

Construction work (such as demolition, site clearance and dewatering) potentially impacts plants and wildlife in the following ways.

- ☑ Disturbance of birds, bats, badgers and other protected species.
- ☑ Removal and fragmentation of habitats.
- ☑ Disturbance to aquatic wildlife and water quality.
- ☑ Disturbance to wildlife from noise and vibration.
- ☑ Damage to trees and hedgerows.
- ☑ Changes in lighting conditions.

Refer to Appendix D for a list of construction work and its potential adverse effects on wildlife.

When making a planning application, developers must produce an ecological impact assessment identifying intended mitigation measures before, during and after construction work.

7.2 Important points

The following practices should be employed to avoid damage to wildlife and its habitats

Before work starts

- ☑ Identify wildlife features and ecologically important areas prior to works commencing and designate protected areas (for example, fence them off). The contract documentation should identify sensitive areas that will require protection and management.
- ☑ Liaise with statutory bodies and local groups to explain any mitigation measures to be used.
- ☑ Where there is a need to take, disturb or relocate protected species, consents should be obtained from the relevant regulatory body and competent licensed ecologists used to carry out the work *(refer to 7.3).*
- ☑ Plan site clearance or demolition works to avoid any nesting, hibernation or breeding seasons. *(Refer to Appendices A and B for a yearly ecology planner that identifies the constraints and best times for dealing with the main groups of protected species.)*
- ☑ Inform and explain to site personnel any protected areas and consequences of any damage to these areas.

During construction

- ☑ Regularly check the condition of fencing of any designated protected areas.
- ☑ Refer to the ecology year planner to be aware of differing seasons and constraints.
- ☑ Ensure watercourses are free from contaminated run-off or any other forms of pollution.
- ☑ Confirm compliance against method statements, the construction environmental management plan and any environmental contractual requirements.
- ☑ In the event of any unexpected ecological finds, stop work and then consult with the site ecologist (if relevant) and statutory bodies.

7.3 Regulatory bodies for nature conservation

Organisations with responsibility for nature conservation in each of the devolved administrations are shown in the table below.

Area	Responsible organisation
England	Natural England
Northern Ireland	Northern Ireland Environment Agency
Scotland	Scottish Natural Heritage
Wales	Natural Resources Wales

In April 2013 Natural Resources Wales (NRW) took over the work of the Countryside Council for Wales, who had joined with the Environment Agency Wales and the Forestry Commission Wales to become a single environment body (SEB).

The organisations' remit includes providing ecological advice, consultation during the planning process, promoting biodiversity, and protecting designated ecological sites and protected species.

There are many other bodies involved in the protection and enhancement of plants and wildlife. The Joint Nature and Conservation Committee (JNCC) is the main public body that advises the UK Government and devolved administrations on UK-wide and international nature conservation. Originally established under the Environmental Protection Act, JNCC was reconstituted by the Natural Environment and Rural Communities (NERC) Act.

JNCC is led by the joint committee, which brings together members from the nature conservation bodies for England, Northern Ireland, Scotland and Wales and independent members appointed by the Secretary of State for the Environment, Food and Rural Affairs under an independent chairperson.

7.4 Endangered species

The natural world slowly changes over thousands of years, during which time new species evolve and some species decline and may ultimately become extinct. Human activity, however, can have a dramatic impact on these natural cycles by putting significant pressure on ecosystems through vegetation clearance, deforestation and habitat fragmentation caused by agriculture, development, road building and other infrastructure projects. This can lead to an accelerated decline in species and the habitats that support them.

The responsible management of wildlife and its habitat is critical to ensure that a healthy, sustainable planet and its renewable resources are passed on to future generations.

At an international level, the International Union for Conservation of Nature (IUCN) is recognised for its work in publishing the *Red list of threatened species*, which is the definitive international standard for species extinction risk. The *Red list* is regularly updated with new information of the status of endangered species.

In the UK, priority species and habitats are those that have been identified as being the most threatened and requiring conservation action under the UK biodiversity action plan (UK BAP). Section 41 of the Natural Environment and Rural Communities Act identifies priority species and habitats at risk of human-induced extinction.

The UK Post Biodiversity framework succeeds the UK BAP and is the result of a change in strategic thinking following the Convention on Biological Diversity's publication *Strategic plan for biodiversity 2011–20.*

The new framework sets out the priorities for UK level work to support the *Strategic plan for biodiversity 2011-20* and its five strategic goals *(shown below)* and 20 Aichi targets (a list of 20 global actions to protect biodiversity).

- ☑ **Strategic goal A.** Address the underlying causes of biodiversity loss by mainstreaming biodiversity across government and society.
- ☑ **Strategic goal B.** Reduce the direct pressures on biodiversity and promote sustainable use.
- ☑ **Strategic goal C.** Improve the status of biodiversity by safeguarding ecosystems, species and genetic diversity.
- ☑ **Strategic goal D.** Enhance the benefits to all from biodiversity and ecosystem services.
- ☑ **Strategic goal E.** Enhance implementation through participatory planning, knowledge management and capacity building.

The UK lists of priority species and habitats remain important and are valuable reference sources. They were used to help draw up statutory lists of priorities in England, Northern Ireland, Scotland and Wales. The most recent list of UK priority species and habitats represents the most comprehensive analysis of such information ever undertaken in the UK. Following this review, the UK priority list now contains **1,149 species** and **67 habitats**. Species were assessed according to four criteria, as shown below.

- ☑ Threatened internationally.
- ☑ International responsibility and a 25% decline in the UK.
- ☑ More than 50% decline in the UK.
- ☑ Other important factors, where data on decline was lacking but there is other evidence of extreme threat.

Examples of **priority species** in the UK are shown below.

Birds	Mammals	Amphibians and reptiles
■ Skylark. ■ House sparrow. ■ Lesser spotted woodpecker.	■ Dormouse. ■ Water vole. ■ Brown long-eared bat. ■ Red squirrel.	■ Common lizard. ■ Great crested newt. ■ Common toad. ■ Adder.
 *Lesser spotted woodpecker*	 *Bats roosting*	 *Great crested newt*

7.5 Protected species and habitats

Several hundred species of birds, wild creatures and plants are protected under the Wildlife and Countryside Act (as amended) and the Wildlife (Northern Ireland) Order (as amended) and are listed in various schedules to the act, for example:

- ☑ Schedule 1: Birds
- ☑ Schedule 5: Animals
- ☑ Schedule 8: Plants.

The Conservation (Natural Habitats, etc.) Regulations (as amended) cover the protection of European protected species – plants and animals listed in the European Habitats Directive.

The habitat regulations include the following.

- ☑ Protection of certain species of animals, Schedule 2 (such as the great crested newt, dormouse, bats, otter and large blue butterfly).
- ☑ Protection of species of plants, Schedule 4 (such as the fen orchid and early gentian).
- ☑ Designation of special areas of conservation (SAC) and special protection areas (SPAs), which are intended to protect the habitats of threatened species of wildlife (see below).

Local planning authorities will, in consultation with Natural England, Natural Resources Wales, Northern Ireland Environment Agency or Scottish Natural Heritage, consider SPAs to protect birds from the effects of disturbance, shooting, egg collecting or other work.

Disturbing protected species or damaging their habitats could result in prosecution against the above legislation. For example, if species such as badgers, bats or great crested newts are disturbed this may result in a fine of £5,000 per offence, which may include the confiscation of plant, equipment and vehicles.

7.5.1 Designated sites

The UK and Northern Ireland have a responsibility to ensure the protection of species and their habitats from both a national and international perspective. One approach to achieving this is to establish designated protected sites, as detailed below.

Areas of outstanding natural beauty (AONB). In England, Northern Ireland and Wales, the primary purpose of the AONB designation is to conserve natural beauty. This, by statute, includes wildlife, physiographic features and cultural heritage, as well as the more conventional concepts of landscape and scenery. Account is taken of the need to safeguard agriculture, forestry and other rural industries and the economic and social needs of local communities. AONB have equivalent status to National Parks as far as conservation is concerned.

AONB are designated under the National Parks and Access to the Countryside Act amended in the Environment Act. The Countryside and Rights of Way Act clarifies the procedure and purpose of designating AONB.

Originally designated in Northern Ireland under the Amenity Lands Act (Northern Ireland), AONB are now designated under the Nature Conservation and Amenity Lands Order (Northern Ireland).

In Scotland, National Scenic Areas are broadly equivalent to AONB.

Areas of special protection for birds (AOSPs) (England and Wales) and **Wildlife refuges** (Northern Ireland) are established under the Wildlife and Countryside Act and the Wildlife (Northern Ireland) Order. The purpose of such orders is normally to provide sanctuary to particularly vulnerable groups of birds. In Scotland these areas of protection are called special protection areas (SPAs).

Biogenetic reserve. A number of national nature reserves (NNR) and some important sites of special scientific interest (SSSI) have been identified as biogenetic reserves under a council of Europe programme for the conservation of heathlands and dry grasslands.

Country parks are statutorily declared and managed by Local Authorities in England and Wales under the Countryside Act and in Scotland under the Countryside (Scotland) Act. In Northern Ireland they exist as a non-statutory designation. Country parks are primarily intended for recreation and leisure opportunities close to population centres and do not necessarily have any nature conservation importance. Nevertheless, many are in areas of semi-natural habitat and so form a valuable network of locations at which informal recreation and the natural environment coexist.

Local nature reserves (LNRs) (England, Scotland and Wales) and **Local Authority Nature Reserves (LANRs)** (Northern Ireland) may be established by Local Authorities under Section 21 of the National Parks and Access to the Countryside Act. These habitats of local significance can make a useful contribution to nature conservation and public amenity.

Marine nature reserves (MNRs) are designated under the Wildlife and Countryside Act (England, Scotland and Wales) to conserve marine flora or fauna, geological or physiographical features, or to allow study of such features. In Northern Ireland they are designated under the Nature Conservation and Amenity Lands (Northern Ireland) Order.

National nature reserves (NNRs) are areas of national, and sometimes international, importance that are owned, leased or approved by the relevant regulatory authority (such as Natural England or Natural Resources Wales), or are managed in accordance with nature reserve agreements between the regulatory authorities and landowners and occupiers. The essential characteristic of NNRs is that they are primarily used for nature conservation.

Ramsar sites. The Ramsar convention requires the protection of wetlands that are of international importance, particularly as waterfowl habitats.

Sites of special scientific interest (SSSI) (England, Scotland and Wales) and **Areas of special scientific interest (ASSI)** (Northern Ireland). All sites of national and international importance on land, including national nature reserves (NNRs), nature conservation review (NCR), geological conservation review (GCR) (England, Scotland and Wales), earth science conservation review (ESCR) (Northern Ireland), special protection areas (SPAs), special areas of conservation (SAC) and Ramsar sites, are notified as SSSI (England, Scotland and Wales) or ASSI (Northern Ireland).

Owners and occupiers are required to notify the relevant regulatory authority of potentially damaging operations and may not undertake them for four months unless they are in accordance with the terms of a management agreement with consent of the relevant regulatory authority.

As protection, the Secretary of State may make a nature conservation order to protect any sites of national or international importance. SSSI are classed as such under the Wildlife and Countryside Act. In Northern Ireland the law relating to ASSI is contained in the Environment (Northern Ireland) Order.

Local planning authorities are required to consult with the statutory bodies (Natural England, Natural Resources Wales, Northern Ireland Environment Agency and Scottish Natural Heritage) prior to allowing any development to proceed that may affect a SSSI or ASSI. Water, gas and electricity companies must also do the same.

Special protection areas (SPAs) and **Special areas of conservation (SAC)** are intended to protect the habitats of threatened species of wildlife. SACs are strictly protected sites under the EC Habitats Directive. This states the requirement to establish a European network of important, high-quality conservation sites.

SPAs are strictly protected sites classified in accordance with Article 4 of the EC directive on the conservation of wild birds (79/409/EEC), also known as the Birds Directive, which came into force in April 1979. They are classified for rare and vulnerable birds, listed in Annex I to the Birds Directive, and for regularly occurring migratory species.

7.5.2 Badgers

The Protection of Badgers Act is provided to protect badgers from deliberate harm, injury or baiting. It is an offence to:

- ☑ disturb a badger when it is occupying a sett
- ☑ interfere with, damage or destroy a sett
- ☑ obstruct access to, or entrance to, a sett
- ☑ wilfully kill, injure, entrap or ill-treat a badger.

A licence must be obtained from the relevant conservation body for any work that may cause disturbance to a badger or involves the damage or destruction of a sett. What constitutes disturbance depends on the nature of the activity proposed. As a rule a licence is normally required for any work within 30 m of a badger sett. However, this distance may increase for more disruptive activities, such as blasting or pile driving.

Licences are not normally issued during the breeding season, which is between 30 November and 1 July, and cannot be issued retrospectively. Activities that necessarily involve disturbance should therefore be programmed to take place outside of this period.

If a badger sett is discovered after works have started, it is essential to stop work immediately and seek expert advice.

Badgers

Artificial badger sett being built as part of mitigation

7.5.3 Great crested newts

Great crested newts are fully protected by law. They rely on water bodies for breeding but otherwise they spend much of their lives on land. They spend winter on land, normally hibernating underground, and emerge soon after the first frost-free days in January or February to begin the migration to breeding ponds (within 500 m).

Initial surveys for great crested newts (by an ecological consultant) are required by the local planning authority and Natural England where water bodies are within 500 m of any proposed development. Four surveys must be made to determine presence or absence, with three surveys conducted between mid-April and mid-May.

Ideally, it is best if great crested newts and their habitats are protected before planning permission has been given for development of a site. If a known or suspected great crested newt site is threatened by a development, the local planning authority and the local office of the statutory nature conservation organisation (SNCO) should be informed as far in advance as possible and the appropriate mitigation scheme approved as part of the planning permission.

Any mitigation proposals will need the submission of a licence application to Natural England, Natural Resources Wales, Northern Ireland Environment Agency or Scottish Natural Heritage. It requires a report that sets out survey results, impacts and a mitigation scheme. The application will take up to 40 days to determine.

e.g. Mitigation for great crested newts

- **Minor impact – on site mitigation.**
  - Small scale relocation and exclusion.
  - Fence erecting and traps set up to exclude newts from works (between March and October), survey for 30 days minimum.
  - Habitat creation (such as creation of refuges and hibernacula – piles of rubble/logs buried beneath surface of ground).
- **Major impact – translocation of newts away from site.**
  - Fence erecting and traps set up (between March and April), survey for 60 days minimum.
  - Translocation of newts to area providing equivalent or better habitat (receptor site).
  - Habitat creation and restoration prior to translocation (such as areas of coarse grassland, hedgerows and ponds).
  - Creation of refuges and hibernaculas (piles of rubble/logs buried beneath surface of ground).

Great crested newt fencing

Great crested newt fencing and mitigation pond

7.5.4 Bats

Bats roost in a number of locations; some examples are shown below.

- ☑ Disused buildings and structures.
- ☑ Under bridges.
- ☑ Holes and cracks in trees.
- ☑ Roofs and walls of buildings.
- ☑ Caves.

Bats hibernate between October and April and breed between May and September. If it is suspected that bats may be present then a bat survey should be carried out by a suitably qualified specialist to establish the size and location of any roost.

Man-made bat house to encourage alternative roosting during construction works

Under the Wildlife and Countryside Act (England, Scotland and Wales) it is illegal to injure, kill, capture or disturb a bat or damage or destroy their roost, even if it is unoccupied at the time. Personnel holding a bat licence are legally allowed to enter a bat roost or to capture, handle or relocate bats. All bats in Northern Ireland are protected under the Wildlife (Northern Ireland) Order and Conservation (Natural Habitats etc.) (Amendment) Regulations (Northern Ireland).

7.5.5 Nesting birds

All birds and their nests are protected in Schedule 1 of the Wildlife and Countryside Act (England, Scotland and Wales) or Section 2 of the Wildlife (Northern Ireland) Order. Some species of bird are further protected from disturbance (such as birds of prey and barn owls).

It is an offence to intentionally kill, injure or take any wild bird. This is backed up by offences of taking, damaging or destroying a nest whilst it is in use or being built, and taking or destroying eggs. In particular, any vegetation clearance works should be done outside of the nesting period.

As highlighted above, it is essential that construction is carefully planned to avoid any works in areas during the nesting period (March to the end of July). Otherwise, any such areas should be fenced off to avoid damage or disturbance. There is also a risk of finding birds nesting in walling stones or pieces of plant during the construction phase. Good awareness is important and site personnel should be made aware of these risks through induction and toolbox talks.

7.5.6 Tree and hedgerow protection

Under Part 8 of the Town and Country Planning Act, the Town and Country Planning (Trees) Regulations and the Planning (Northern Ireland) Order, tree preservation orders (TPOs) can be made by the local planning authority to prohibit the cutting down, uprooting, pollarding, topping, lopping, wilful damage or destruction of trees without the local planning authority's consent. It is illegal to cut down or alter any tree under a TPO. All trees within a designated conservation area are protected.

The Local Authority Planning Department should be contacted before you work on any protected trees. The procedure to gain permission to work on a tree with a TPO is shown below.

- ☑ Obtain the relevant form from the Local Authority.
- ☑ Fill out the form outlining the intended work.
- ☑ Return the completed form to the Local Authority.

The Local Authority then has up to two months to make a decision.

Under the Hedgerows Regulations (England and Wales only), a hedgerow removal notice is required from the Local Authority if it is older than 30 years or satisfies at least one of the criteria listed in Part 2 of Schedule 1. The removal of a hedgerow longer than 20 m requires planning permission.

During the planning of a project, you should make contact with the Local Authority to ensure that any trees or hedgerows identified for removal are discussed and, if possible, avoided. Translocation of mature or ancient hedgerows should also be considered. Works on trees and hedgerows are confined to certain time constraints due to nesting birds and bats, which are protected species.

All trees should be protected in accordance with *Trees in relation to construction* (BS 5837:2012)

The following practices should be followed.

- **Protected zone**: a fenced-off area around the tree to be retained should be established to eliminate tree damage during development. The size of the protected zone is either the distance of the crown spread of the tree (the width of the tree at its widest) or half the height of the tree, whichever is the greater.

Tree and hedgerow protection – good tree protection measures

- **Compaction of soil** must be avoided as this destroys the soil's pore structure, which in turn alters the trees' ability to take up water. Compaction can be caused by the storage of materials, machinery and soil and by the use of an area as a thoroughfare by people and vehicles and for temporary constructions (such as site huts).
- **Excavation** for foundations and utilities must be considered in relation to the welfare of root systems. It is a common misconception that the root system of a tree extends in a narrow, deep band directly below the trunk. In reality the roots extend a large distance laterally at a shallow depth, often at approximately 60 cm. Shallow foundations thus impact on the root systems.

  The closer the excavation is to the tree the greater the chance of damage. Construction, therefore, should not be within the protected zone of the tree. This also benefits the building as roots can damage foundations. Where foundations are to be located near major roots they should carefully bridge them.

  Construction also has an impact on the hydrological cycle with compaction of soils and the laying of tarmac reducing infiltration rates and increasing run-off. This affects water available for tree growth.
- **Ground level change** should acknowledge the location of tree roots. Reduction in ground level causes the severance of roots and an alteration of drainage rates that affect water availability. Increases in ground level cause compaction and can suffocate and damage shallow and fragile roots.
- **Impact** by machinery can damage trees in an obvious way; torn branches, damaged bark and general trunk wounds allow easier access to disease and parasites that cause decay.
- **Contamination of soil** through the leakage of chemicals from construction materials (such as concrete, fuel and oil) should be avoided through secure storage. Storage of these materials must not occur in the protected zone and should be downhill from the tree, ideally 10 m from the protected zone to allow for leaching of materials through the soil.
- **Tree surgery** may require consent if the tree is protected or is within a conservation area. Any surgery or necessary felling should be detailed in the planning application and should be undertaken prior to commencement of construction; it is easier to carry out surgery without unavoidable construction obstacles. It should also be noted that restrictions on felling and surgery could be implemented during bird nesting season.

The National Joint Utilities Group (NJUG) has also published *Guidelines for the planning, installation and maintenance of utility apparatus in proximity to trees (Issue 2)*. The document provides details for the establishment of a tree protection zone and precautions that should be taken for any works within it.

Tree damage caused by inappropriate storage of construction waste within the protection zone

Tree protection zone

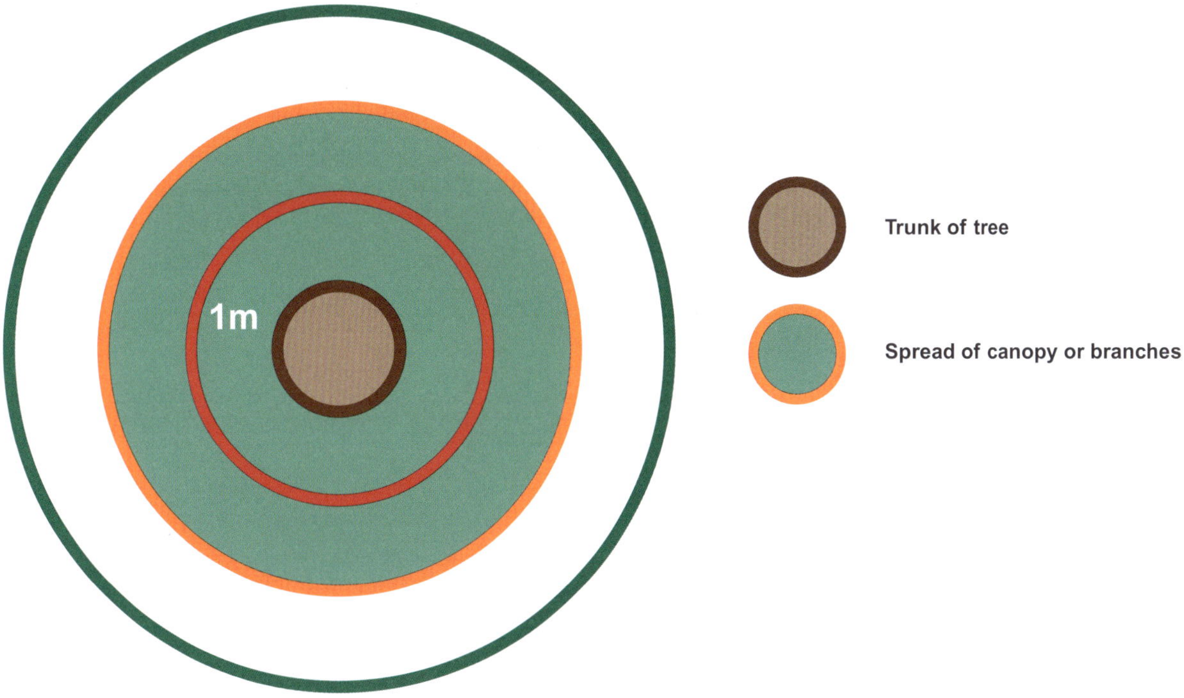

Prohibited zone – one metre from trunk
Excavations of any kind must not be undertaken within this zone, unless full consideration with the Local Authority tree officer is undertaken. Materials, plant and spoil must not be stored within this zone.

Precautionary zone – four times tree circumference
Where excavations must be undertaken within this zone the use of mechanical excavation plant should be prohibited. Precautions should be undertaken to protect any exposed roots. Materials, plant and spoil should not be stored within this zone. Consult with the Local Authority tree officer if in any doubt.

Permitted zone – outside of precautionary zone
Excavation works may be undertaken within this zone. However, caution must be applied and the use of mechanical plant limited. Any exposed roots should be protected.

www. **For further details of these guidelines refer to the NJUG website.**

For an ecology checklist refer to Appendix E.

7.6 Invasive species

Species of plants and animals that have been introduced where they do not naturally occur are known as *non-native species*. A number of these species have become invasive, because they are bigger, faster growing or more aggressive than native species, which are unable to compete. Invasive species therefore have a damaging impact on British plants, animals and ecosystems by spreading disease, competing for habitat and direct predation.

The contract documentation or environmental statement for the project, provided by the client, will highlight any non-native species of concern. If these are present then advice should be sought from specialists, who will provide further details on how best to treat and dispose of them in each instance.

The Environment Agency promotes a *check, clean, dry* initiative. This will help to stop the spread of invasive species (by transfer through equipment, shoes and clothing) from one body of water to another.

The non-native species secretariat (NNSS) has responsibility for helping to co-ordinate the approach to invasive non-native species in Great Britain.

www. **For information on how to stop the spread of invasive species and the *check, clean, dry* campaign refer to the GB non-native species secretariat.**

Relevant control measures for any invasive species associated with the project should be included within the project construction environmental management plan, together with relevant method statements that may need to be agreed with the relevant agencies before commencement of work.

It is possible that construction work has an impact on invasive species of animal, most likely caused by accidental dispersion. It is an offence under Section 14 of the Wildlife and Countryside Act (England, Scotland or Wales) or under Article 15 of the Wildlife (Northern Ireland) Order (as amended) to deliberately permit the spread of an invasive species by releasing it into the wild.

One such example species is the American signal crayfish, which is driving the native white-clawed crayfish towards extinction and causing declines in the diversity and richness of aquatic communities. Commercial fisheries have been affected by predation of fish eggs and competition between crayfish and salmon species for refuges. Burrowing by crayfish can cause erosion of riverbanks and destabilise structures built at the edges of rivers.

Details of invasive species are available on the Government website, including a leaflet on how to manage non-native species.

The **PlantTracker** project is a collaboration between the Environment Agency, Scottish Natural Heritage, the Scottish Environment Protection Agency and Natural Resources Wales. It is part of the nature locator project, which is a collection of projects crowd-sourcing data on biodiversity using smartphone apps and accompanying websites.

The main aim is to locate incidences of a number of high priority invasive plant species. There is currently a lack of information on exactly how serious the problem presented by invasive plant species really is.

Obtaining accurate data about the distribution of invasive species is of paramount importance when it comes to assessing impact and formulating a response.

To download the free app visit the *PlantTracker* website.

Section 14 of the Wildlife and Countryside Act and Article 15 of the Wildlife (Northern Ireland) Order make it an offence to plant or otherwise cause to grow in the wild any of the plants listed in Schedule 9.

- Giant hogweed.
- Giant kelp.
- Japanese knotweed.
- Japanese seaweed.
- Himalayan balsam.
- Rhododendron.
- Floating pennywort.
- Parrot's feather.
- Australian swamp stonecrop.

The most common of these are giant hogweed, Japanese knotweed and Himalayan balsam.

Refer to Appendix C for details on how to deal with Japanese knotweed.

Giant hogweed

Japanese knotweed
(Image supplied by Bridget Plowright)

Himalayan balsam

There are other pieces of legislation that relate to preventing the spread of invasive species in the UK, which can include both native and non-native species (for example, the native species ragwort is one of the species controlled under the Weeds Act).

The following practices should be employed for the effective management of invasive plants

Before work starts

- ☑ Review the environmental statement and conduct a site survey to identify any non-native species.
- ☑ Liaise with landowners and statutory bodies on the appropriate treatment techniques where invasive species have been identified.
- ☑ Demarcate and fence off any areas found to contain non-native invasive species.

During construction

- ☑ Ensure the workforce are made aware of problems with non-native invasive species highlighted at site inductions, using maps of the site, and reinforce this information through the use of toolbox talks.
- ☑ Implement bio-security measures to ensure the work does not contribute to the spread of the non-native invasive species.
- ☑ Maintain fenced areas to restrict access and to prevent spread across the site.
- ☑ Any use of herbicides near or in watercourses needs to be approved by the environment agencies. Ensure that treatment companies hold appropriate certificates of competence.
- ☑ Ensure that any chemical containers or material contaminated with herbicides are disposed of in accordance with the duty of care and Hazardous Waste Regulations as appropriate.
- ☑ Ensure that soils contaminated with invasive plants or roots are removed to an appropriately authorised landfill site.

The management of invasive species will often involve the generation of waste in the form of plant material, soil, water and sediment, or dead animals. This waste needs to be dealt with carefully to ensure that it complies with the relevant waste legislation. For example, when Japanese knotweed is dug up to eradicate it from a site, the soil will contain rhizome (underground root-like stems) from which the plant can regenerate.

This material is likely to be classified as controlled waste under the Environmental Protection Act and can only be transferred by an authorised person (such as a registered waste carrier) and taken to an appropriately authorised landfill site.

For further information on the management of waste refer to Chapter E03 Waste management.

Pesticides Act and Control of Pesticides Regulations

In many cases the management of invasive plants will involve the use of herbicides, which are controlled under the Pesticides Act and the Control of Pesticides Regulations. These pieces of legislation place strict controls on the supply, storage and use of all pesticides, including those herbicides used in the control of invasive species. These regulations make it obligatory for those giving advice on the use of pesticide products to hold a certificate of competence recognised by DEFRA. The only qualifications recognised by DEFRA are those issued by the British Agrochemical Supply Industry Scheme (BASIS). Permission is also required from the relevant environment agency if spraying in and near watercourses.

7.7 Promoting biodiversity

Working together, planners and developers or clients should, through good design, aim to minimise the ecological impact of the project (such as habitat destruction, fragmentation and species isolation) and, wherever possible, should actively seek to rebuild local ecological networks. This can be achieved through the points set out below.

- ☑ **Enhance** the overall ecological quality of the site and the surrounding ecological network by creating new habitats, buffer areas and landscape features that are of importance for wildlife. Focus should be made:
  - in areas where the most important, fragile and/or threatened habitats and species are known to occur
  - where there are species requiring large ranges and/or those with limited powers of dispersal, which have particularly suffered as a result of habitat becoming reduced in size and isolated
  - on species with low reproductive capacity (most large mammals) or species highly sensitive to disturbance (most birds of prey) and species subject to recovery programmes (the focus of local biodiversity action plan targets).
- ☑ **Local biodiversity action plans**, species action plans and habitat action plans should be used as a guide to the relevant priorities for such positive measures at the local level. These may include rehabilitation of degraded habitats or the creation of new habitats within and adjacent to development sites.
  - Bird boxes, swift bricks, bat boxes, bat bricks, green roofs, green walls and habitat creation are examples where enhancement can be achieved on construction projects.

- ☑ **Avoid** developing sites, and locations within sites, where existing habitats, important species, buffer areas and other landscape features of major importance for wildlife would:
  - suffer direct impact resulting in the reduction or complete loss of habitat and/or diversity of species present
  - suffer an indirect impact from nearby development through increased ecological disturbance and stress, thereby reducing the site's capacity to support the wildlife present
  - suffer a reduction in ecological quality so that the site is no longer able to support the migration, dispersal or genetic exchange of wild species
  - be further fragmented from other similar features by development that causes a barrier effect in the landscape between fragments.
- ☑ **Restore** and, where possible, link and connect existing habitats and landscape features, which could potentially be of major importance for wildlife, enhancing their ability to support migration, dispersal and genetic exchange.
- ☑ **Retain and incorporate** within the development site layout existing habitats, important species, buffer areas and landscape features of major importance for wildlife – making sure that the site retains at least the same capacity to support the diversity, abundance, migration, dispersal and genetic exchange of wildlife as it did prior to development.
- ☑ **Compensate** for features lost to development through the following.
  - Re-creation, as nearby as possible, of features and landforms capable of maintaining the same quality of habitats and species as would otherwise be lost or displaced through the development.
  - Restoration and enhancement of surrounding or nearby features unaffected by development.
  - Creation of new or additional buffer areas to reduce impacts.
  - Translocation, where possible, of habitats and species that would otherwise be lost.

www.

☑ For further information on local biodiversity action plans refer to the Business and Biodiversity Resource Centre's website.

☑ Additional resources can be found at CIRIA in its publication *Working with wildlife: guidance for the construction industry* (C691).

Appendix A – Wildlife year planner

Traditionally, environmental issues are not considered at an early stage in project planning. The year planners below show the appropriate times of year to deal with specific issues.

Badgers

Task	Jan	Feb	Mar	Apr	May	June	July	Aug	Sept	Oct	Nov	Dec
1	*	*	*	*							*	*
2					*	*	*	*	*	*		
3	*	*										*
4		*	*	*								
5							*	*	*	*	*	

1. **Badger surveys** are best carried out during November to April when the vegetation is low and field signs are easier to identify.
2. **Badger surveys** can continue but become less reliable as the vegetation becomes denser.
3. **Artificial badger sett construction** should be completed at least six months before a sett exclusion. Winter is the time to complete artificial setts to give the badgers time to familiarise themselves before the licensing season.
4. **Badger territorial bait marking surveys.** Territorial marking is at its peak between February and April and the vegetation is low enough to identify latrines. This is the only time of year when bait marking is effective.
5. **Badger licensing season: Between 1 July and 30 November.** Nature conservancy councils will normally only issue disturbance and exclusion licences outside this time period in cases of proven urgency.

Water voles

Task	Jan	Feb	Mar	Apr	May	June	July	Aug	Sept	Oct	Nov	Dec
1			*	*	*	*						
2							*	*	*	*		
3			*	*						*	*	

1. **Water vole surveys** need to be carried out during the summer breeding season. Best carried out between March and June before the vegetation grows too high.
2. **Water vole surveys** can continue for the rest of the breeding season but field signs become harder to identify as the vegetation becomes denser.
3. **Water vole exclusions.** Excluding and trapping are recommended during February and March, before the breeding season, or October and November, after the end of the breeding season.

E
07

Otters

Task	Jan	Feb	Mar	Apr	May	June	July	Aug	Sept	Oct	Nov	Dec
1	*	*	*	*	*	*	*	*	*	*	*	*
2	*	*	*	*	*	*	*	*	*	*	*	*

1. **Otter surveys** can be carried out throughout the year, although surveys are easier to carry out during periods of low vegetation.
2. **Otter mitigation** can be carried out throughout the year but may be restricted if evidence of breeding is identified.

Great crested newts

Task	Jan	Feb	Mar	Apr	May	June	July	Aug	Sept	Oct	Nov	Dec
1			*	*	*	*						
2		*	*				*	*	*	*	*	

1. **Great crested newts breeding pond surveys** are the only way to effectively establish presence or absence and to quantify populations. These are best carried out between March and June.
2. **Terrestrial searches** are least effective but can be carried out during early spring and during the autumn, depending upon weather conditions. Surveys should be carried out when night temperatures are above five degrees centigrade and when the ground is moist.

Bats

Task	Jan	Feb	Mar	Apr	May	June	July	Aug	Sept	Oct	Nov	Dec
1			*	*	*	*	*	*	*	*		
2					*	*	*	*	*	*		
3	*	*										*
4				*	*	*	*	*	*	*		

1. **Flight surveys** that involve identification of bats in flight by observation or echolocation can be carried out between March and October, although the optimum time for the surveys is April to September. Surveys of this type can be carried out without a licence as they are non-intrusive.
2. **Dusk emergence and dawn swarming surveys** can be carried out between May and October although the optimum time for the surveys is between May and September.
3. **Hibernation roost surveys** can only be carried out between November and March when the bats are hibernating. These surveys are very difficult to carry out as bats hibernate deep in cracks and crevices and are therefore difficult to identify.
4. **Habitat surveys** can only be effectively carried out between April and October when the bats are active.

Crayfish

Task	Jan	Feb	Mar	Apr	May	June	July	Aug	Sept	Oct	Nov	Dec
1				*			*	*	*	*		
2					*	*						
3	*	*	*								*	*

1. **Crayfish surveys** are best carried out between July and October, although it is possible to carry out surveys during April.
2. **Crayfish releasing young surveys** should not be carried out during May and June because crayfish could still be carrying their newly hatched young.
3. **Crayfish reduced activity surveys** should not be carried out during the winter months due to reduced levels of activity.

Nesting birds

Task	Jan	Feb	Mar	Apr	May	June	July	Aug	Sept	Oct	Nov	Dec
1			*	*	*	*	*	*	*			
2	*	*								*	*	*

1. **Nesting bird season.** No vegetation clearance work should be carried out during the nesting bird season unless immediately preceded by a thorough nesting bird survey.
2. **Vegetation clearance work** is best carried out at these times of year when birds are not nesting, although work must stop if nests are found.

Reptiles

Task	Jan	Feb	Mar	Apr	May	June	July	Aug	Sept	Oct	Nov	Dec
1			*	*	*	*	*	*	*	*		
2			*	*	*	*	*	*	*	*		
3	*	*	*	*	*	*	*	*	*	*	*	*

1. **Reptile surveys** can only be carried out between March and October when reptiles are active. During the winter months reptiles are in hibernation, therefore it is not suitable to carry out surveys.
2. **Reptile capture and release programmes** can only be carried out between March and October when reptiles are active.
3. **Scrub clearance** work can be carried out throughout the year.

Appendix B – Wildlife year planner guidance

The following tables provide guidance on the optimal timing for carrying out specialist ecological surveys and mitigation.

		Licence required?	J	F	M	A	M	J	J	A	S	O	N	D
Habitats/vegetation	Surveys	N	Mosses and lichens. No other detailed plant surveys – Phase 1 only (least suitable time)		Detailed habitat assessment surveys. Surveys for higher plants and ferns. Mosses and lichens in April, May and September only								Mosses and lichens No other detailed plant surveys – Phase 1 only (least suitable time)	
	Mitigation	N	Planting and translocation		No mitigation for majority of species								Planting and translocation	
Badgers	Surveys	N**	Optimal time during spring and autumn when activity levels are high, vegetation levels low and field signs most obvious											
	Mitigation	***	No disturbance of existing setts. Building of artificial setts						Stopping up and excavation of existing setts Licence may be required					See Jan-June
Bats	Surveys	*	Hibernation surveys and inspection of building and tree roosts		Inspection of building and tree roosts	Optimal time for emergence/re-entry surveys from May to August. Other activity surveys may extend from April to October. Inspection of tree and building roosts.							Hibernation surveys & inspection of building and tree roosts	
	Mitigation	***	Works on non-breeding summer and maternity roosts only				Works on hibernation roosts				Works on non-breeding summer and hibernation roosts	Works on non-breeding summer, maternity and hibernation roosts	Works on non-breeding summer and maternity roosts only	
Birds	Surveys	*	Winter birds		Breeding birds and migrant species						Migrant species August through to October		Winter birds	
	Mitigation	N	Clearance works possible but must stop immediately if any nesting birds found		Avoid clearance or construction works in nesting habitat or carry out under Ecological Watching Brief						Clearance works possible but must stop immediately if nesting birds found			
Dormice (n/a in NI)	Surveys	*	Nut searches	Nut searches (sub-optimal time)		Nest tube surveys April–November. Nest searches – any time of year but preferably September to March. Nut searches (September-November)								Nut searches
	Mitigation	***	No clearance works				Clearance works (sub-optimal time)	No clearance works			Clearance works to early October (optimal time)		No clearance works	

Key

(dark green)	Recommended survey period	(light orange)	Recommended timing for mitigation works
(pale blue)	No surveys	(dark orange)	Mitigation works restricted

E 07

		Licence required?	J	F	M	A	M	J	J	A	S	O	N	D
Fish	Surveys	*	Timing of surveys may be dependent on the migration pattern of the species concerned and the breeding season of individual fish species											
	Mitigation	***	Mitigation for the protection of watercourses is required at all times of year. Mitigation should be timed to avoid fish breeding season which will vary from species to species											
Great crested newts (n/a in NI)	Surveys	*	No surveys – newts in hibernation		Pond surveys for adults: mid-March to mid-June. Surveys must include visits between mid-April and mid-May. Egg searches April to mid-June. Larval surveys from mid-May. Terrestrial habitat surveys				Larval surveys to mid-August. Terrestrial habitat surveys		Terrestrial habitat surveys		No surveys – newts in hibernation	
	Mitigation	***	No trapping of newts. Pond management only		Newt trapping programmes in ponds and on land. Note trapping may be limited by cold night temperatures in early months				Newt trapping on land only. Note trapping may be limited by dry nights during July and August				No trapping of newts. Pond management only	
Natterjack toads (n/a in NI)	Surveys	*	No surveys – toads in hibernation			Surveys of breeding ponds for adults April–June. Surveys for tadpoles from May onwards.				Surveys for adults on land		No surveys – toads in hibernation		
	Mitigation	***	Pond management works			Trapping of adults in ponds from April to June. Trapping on adults on land April to September. Trapping of tadpoles from May to early September						Pond management works		
Reptiles: adder, grass snake slow worm and common lizard (n/a in NI except common lizard)	Surveys	N	No surveys – reptiles in hibernation		Activity surveys from March to June and in September/October. Note surveys are limited by high temperatures during July and August. Peak survey months are April, May and September								No surveys – reptiles in hibernation	
	Mitigation	N	Scrub clearance – proceed with caution to avoid disturbance of hibernating reptiles		Capture and translocation programmes can only be conducted while reptiles are active (March–June and September/October). Note trapping is limited by high temperatures during July and August								Scrub clearance – proceed with caution to avoid disturbance of hibernating reptiles	
Common lizard (NI only): sand lizard, smooth snake	Surveys	*	No surveys – reptiles in hibernation		Activity surveys possible from March to June and in September/October but peak survey months are April, May and September. Surveys are limited by high temperatures during July and August.								No surveys – reptiles in hibernation	
	Mitigation	***	Scrub clearance – proceed with caution to avoid disturbance of hibernating reptiles		Capture and translocation programmes can only be conducted while reptiles are active (March–June and September/October). Note trapping is limited by high temperatures during July and August.								No clearance works – proceed with caution to avoid disturbance of hibernating reptiles	
Otters	Surveys	N**	Surveys possible all year round											
	Mitigation	***	Mitigation possible all year round but timing will be restricted where otters are breeding											

<table>
<tr><th></th><th></th><th>Licence required?</th><th>J</th><th>F</th><th>M</th><th>A</th><th>M</th><th>J</th><th>J</th><th>A</th><th>S</th><th>O</th><th>N</th><th>D</th></tr>
<tr><td rowspan="2">Pine marten</td><td>Surveys</td><td>N**</td><td colspan="12">Surveys possible all year round but optimal time is during spring and summer. Breeding den surveys between March and May</td></tr>
<tr><td>Mitigation</td><td>***</td><td colspan="2">Works in pine marten habitat</td><td colspan="8">Avoid works in pine marten habitat</td><td colspan="2">Works in pine marten habitat</td></tr>
<tr><td rowspan="2">Red squirrel</td><td>Surveys</td><td>*</td><td colspan="12">Surveys possible all year round but optimal time is during spring and summer.
Surveys for breeding females from January to September, with some surveys during peak breeding periods of March–May and/or July–September.</td></tr>
<tr><td>Mitigation</td><td>***</td><td colspan="9">Avoid all works in red squirrel habitat</td><td colspan="2">Works should be conducted at this time</td><td>See Jan-Sept</td></tr>
<tr><td rowspan="2">Smooth newt (NI only)</td><td>Surveys</td><td>*</td><td colspan="2">No surveys – newts in hibernation</td><td colspan="4">Pond surveys for adults: March to mid-June. Surveys should include visits between mid-April and mid-May. Egg searches April to mid-June. Larval surveys from mid-May. Terrestrial habitat surveys</td><td colspan="2">Larval surveys to mid-August.
Terrestrial habitat surveys</td><td colspan="2">Terrestrial habitat surveys</td><td colspan="2">No surveys – newts in hibernation</td></tr>
<tr><td>Mitigation</td><td>***</td><td colspan="2">No trapping of newts. Pond management only</td><td colspan="4">Newt trapping programmes in ponds and on land. Note trapping may be limited by cold night temperatures in early months</td><td colspan="4">Newt trapping on land only. Note trapping may be limited by dry nights during July and August</td><td colspan="2">No trapping of newts. Pond management only</td></tr>
<tr><td rowspan="2">Water vole (n/a in NI)</td><td>Surveys</td><td>N**</td><td colspan="2">Reduced activity</td><td colspan="8">Activity and breeding surveys depending on vegetation cover and weather conditions.
Optimum survey period March–June</td><td colspan="2">Reduced activity</td></tr>
<tr><td>Mitigation</td><td>***</td><td colspan="2">Avoid all works in water vole habitat</td><td colspan="2">Works in water vole habitat possible</td><td colspan="3">Avoid all works in water vole habitat</td><td colspan="2">Works in water vole habitat possible</td><td colspan="3">Avoid all works in water vole habitat</td></tr>
<tr><td rowspan="2">White-clawed crayfish</td><td>Surveys</td><td>*</td><td colspan="3">Reduced activity</td><td colspan="2">Surveys possible</td><td colspan="2">Avoid surveys (females are releasing young)</td><td colspan="3">Optimum time for surveys</td><td colspan="2">Reduced activity</td></tr>
<tr><td>Mitigation</td><td>***</td><td colspan="3">Avoid capture programmes (low activity levels may mean animals are missed)</td><td>Exclusion from construction areas</td><td colspan="2">Avoid capture programmes – breeding period</td><td colspan="4">Exclusion from construction areas</td><td colspan="2">Avoid capture programmes (low activity levels may mean animals are missed)</td></tr>
</table>

* accepted survey and monitoring techniques may involve the capture, handling or disturbance of these protected species (in the case of birds, those listed on Schedule 1 of the WCA only). Where this is the case, only licensed persons can undertake the surveys. These are obtained from Natural England, Countryside Council for Wales, Northern Ireland Environment Agency or Scottish Natural Heritage

** accepted survey and monitoring techniques do not typically involve the capture, handling or disturbance of these protected species and so a survey licence is not ordinarily required. However, should further techniques be used that will result in the above, only licensed persons can undertake the survey

*** where mitigation involves the capture, handling or disturbance of a protected species and/or the damage, destruction or obstruction of their habitats, a conservation or mitigation licence must be obtained from Natural England, Countryside Council for Wales, Northern Ireland Environment Agency or Scottish Natural Heritage. Mitigation licence applications take about 30 working days to be processed by government agencies. Where mitigation works need to be conducted under licence before works begin, licence applications need to be submitted considerably earlier.

Appendix C – Dealing with Japanese knotweed

Japanese knotweed can cause structural damage. It can grow up to 20 mm a day and is strong enough to penetrate foundations. It is an offence under the Wildlife and Countryside Act (England, Scotland and Wales) and the Wildlife (Northern Ireland) Order (as amended) to encourage the spread of Japanese knotweed. It is often found along railways, riverbanks, roads and derelict sites. Any known areas of Japanese knotweed will be highlighted within the environmental statement for the project. These areas must be cordoned off to prevent inadvertent spread. Extreme care must be taken to ensure that all equipment used on site is free of Japanese knotweed material before leaving the site to avoid contamination.

If works are required in areas of known Japanese knotweed strands, control methods must be applied; the particular method to be employed will be dependent on site conditions. Where possible, excavation should be avoided and the plant should be treated in its original position.

(Image supplied by Bridget Plowright)

The Japanese knotweed rhizome system (root zone) may extend to, and beyond, a depth of at least 2 m and extend 7 m laterally from a parent plant.

Chemical control

Chemical control usually takes a minimum of three years to totally eradicate Japanese knotweed. Wherever there is a risk of contamination to a watercourse, choice of herbicide is limited to formulations of Glyphosate and 2, 4-D Amine that are approved for use in or near water. Use of herbicides in or near water requires formal consultation with the relevant environment agency. Spraying both the top and underside of leaves improves control. Plants respond best when actively growing. The most effective time to apply herbicide is from July to September.

Non-chemical controls

Cutting. Use a simple scythe method of cutting to prevent stem fragmentation. Continue cutting every 2–4 weeks to reduce both above and below-ground biomass. (This will only control the knotweed, not eradicate it.)

Burning. Controlled burning of stem and crown material may be used as part of the control programme. Such burning must take into account the potential for nuisance or pollution that may occur as a result of the activity. Burning in the open may be undertaken in accordance with a registered exemption in accordance with the Environmental Permitting Regulations.

Burial. Soil containing knotweed material and burnt remains of knotweed may be buried on site of production. On-site burial must be performed to a depth of at least 5 m. Knotweed material should be covered with root barrier membrane consisting of a geotextile layer or a heavy gauge polythene sheet prior to infilling. It is strongly advised to record the burial site location, and to inform any future owners of its position.

Excavation. Only consider excavating as a last resort, unless this is part of an on-site treatment method. When using excavation for off-site disposal, take great care to avoid excess waste and make sure the excavated material does not contaminate surplus soil.

Off-site disposal. Where the option for on-site disposal or treatment is not available, as a last resort in terms of sustainability, material contaminated with Japanese knotweed must be disposed of at an approved disposal facility, having informed the site operator of the presence of viable knotweed within the material.

To ensure compliance with the Wildlife and Countryside Act and to reduce the risk of spreading knotweed, any such on-site burial and any controlled burning must be done in accordance with the Environment Agency's knotweed Code of Practice *Managing Japanese knotweed on development sites.*

A copy of the Environment Agency's Code of Practice for managing Japanese knotweed can be obtained from its website.

Failure to appropriately dispose of any material containing Japanese knotweed may lead to prosecution under Section 33 and 34 of the Environmental Protection Act and Section 14 of the Wildlife and Countryside Act.

Appendix D – Construction work and its potential adverse effects on wildlife

Construction work and its potential adverse effects on wildlife		
Construction activity	**Implication**	**Examples of effect on wildlife**
Site clearance	Removal of trees and shrubs	▪ Loss of important species or specimens of tree or shrub that may be protected by a tree preservation order (TPO) ▪ Loss of bird nests or bat roosts ▪ Loss of habitat for protected species ▪ Loss of important invertebrates, including those that may require deadwood habitat (such as stag beetles)
	Removal of ground vegetation	▪ Loss of habitat for protected species ▪ Loss of rare plants ▪ Loss of bird nests ▪ Killing or injury of reptiles or amphibians ▪ Killing or injury of small mammals ▪ Loss of invertebrates and their breeding habitat
	Removal of soil	▪ Loss of habitat for protected species ▪ Loss of seed bank ▪ Loss of water vole burrows ▪ Loss of invertebrates and their breeding habitat ▪ Destruction of badger setts
	Demolition of buildings and structures	▪ Loss of bird nesting or bat roosting areas
	Removal of rubble and other materials	▪ Loss of reptile and amphibian habitat
Site set up	Location of site offices and compounds	▪ Disturbance of breeding animals
	Storage areas	▪ Potential pollution of important watercourses, wetlands or other water bodies, including coastal waters, through spillage or dust
Establishment of haul roads	Rubble or concrete temporary roads constructed	▪ Fragmentation of habitats ▪ Road kills ▪ Destruction of badger setts ▪ Contamination of adjoining habitats by dust ▪ Noise or light pollution may disturb nesting birds or other animals ▪ Change of pH through leaching
Groundworks	Ground investigations Foundations Excavations and piling Temporary earthworks Tunnelling	▪ Impacts on surface and groundwater, which may have secondary impacts on important wetlands both on and off site ▪ Noise or light pollution may disturb nesting birds or other animals ▪ Destruction of badger setts ▪ Run-off and erosion, which may damage important habitats ▪ Potential to introduce or spread invasive plants (such as Japanese knotweed)
Construction	Concrete pours and other wet trades	▪ Contamination of wetlands ▪ Change of soil pH through run-off

Appendix E – Ecology checklist

Company name		Project title	
Location		Contract no.	

Ecology

	Yes	No	N/A
1. Has the contractual and client documentation been reviewed to identify all sensitive areas containing wildlife (animals and vegetation)?			
2. Have all relevant ecological risks been included in the construction environmental management plan, together with their mitigation measures?			
3. Have licences been obtained from the relevant conservation body to move protected species or disrupt their habitats (such as bats, badgers and lizards)?			
4. Have specific bio-security plans been worked up and implemented if non-native species have been identified?			
5. Are areas containing wildlife suitably protected from the construction work?			
6. Are trees suitably protected to avoid damage from the works?			
7. Have vegetation removal and other construction works been programmed to avoid breeding and nesting periods?			
8. Has permission been given by the Local Authority to remove hedges or trees that have a tree preservation order (TPO)?			
9. Has approval been given by the relevant environment agency to deal with invasive plants (such as Japanese knotweed and giant hogweed)?			
10. Has approval been given by the relevant environment agency for the treatment or management of plants adjacent to a watercourse?			
11. Do treatment operatives hold the appropriate qualification and certificate of competency for the use of pesticides?			

Comments

Name		Position		Signature		Date	

For an interactive PDF version of this form visit the companion website.

08 Contaminated land

Contents

Supporting information

G: Checklists and forms

GT 700 Toolbox talks

Overview

This chapter gives a general overview of the legal framework for the definition, regulation and management of contaminated land.

The chapter identifies the process for assessing and remediating contaminated land, the licences that will be required for its treatment or disposal and guidance for the prevention of pollution. It will also identify the main occupational health considerations when dealing with contaminated land.

8.1 Introduction

Building and construction works often involve the redevelopment of land that was used previously for commercial or industrial activities. These sites are often referred to as brownfield sites, and bringing them back into use is considered a more sustainable option than using new or greenfield land. However, the UK's extensive industrial past has left a legacy of contamination on many sites, which needs to be dealt with before they can be suitable for their new purpose.

Both the surface of the ground and the ground beneath the surface may be contaminated by materials that have been worked, stored, spilt, buried, dumped or abandoned on the land in previous years. This list will also include the residue, waste or by-products from some industrial processes and the ashes from fires. Both solid and liquid waste may have permeated the ground to a considerable depth.

Sites that have had previous industrial occupation should be assumed to be polluted, and tests undertaken to ascertain the types of pollutant and their levels of concentration.

Everyone involved in work on such land must be made aware of the contamination. They must make an assessment of potential risks to human health and the environment, and implement any protective measures that need to be taken.

8.2 Important points

- ☑ The contract documentation and planning conditions for a project will identify known contamination of the site and the agreed methods for dealing with it. These should be referred to, and their requirements included, in the construction environmental management plan.
- ☑ Where a contaminated land assessment needs to be undertaken, this will be carried out in accordance with a systematic process of investigation, testing and appraisal to identify the most appropriate method for its treatment. The agreed method should be approved by the regulators before works commence.
- ☑ The testing of contaminated land must be carried out by certified, competent staff in accordance with standard field testing and laboratory procedures approved by the regulators.
- ☑ All contaminated areas of the site must be clearly signed and fenced off to avoid unauthorised access and inadvertent spread of contamination across the site.
- ☑ The use of mobile plant for remediation of contaminated soils will require an environmental permit for the equipment and the need to complete a site deployment form detailing the work and management of the risks at each specific location.
- ☑ Where, following treatment, materials are still classed as waste, then their use will also require an environmental permit or registered exemption. The Contaminated Land: Applications in Real Environments (CL:AIRE) development industry Code of Practice can be used to ensure that contaminated materials are treated and managed so that they achieve an end of waste status. This will require the implementation of a materials management plan and a declaration by a qualified person that all requirements have been complied with.
- ☑ The stockpiling of contaminated soils should be avoided. Where stockpiling is unavoidable, material should be placed on impervious ground to avoid contaminates seeping into the ground and it should be covered to avoid wind-blown contamination or run-off to drainage systems and watercourses.
- ☑ Any removal of waste off site for treatment elsewhere or disposal should comply with the duty of care, including the provision of waste transfer or consignment documentation.
- ☑ All vehicles carrying contaminated materials off site should be appropriately sheeted and should pass through a wheel-wash facility to avoid contamination of the public highway.
- ☑ All personnel involved in the treatment of contaminated land should be aware of the relevant risks associated with the particular contaminants and wear protective clothing, gloves and boots. Depending on the level of risk, a decontamination unit may need to be employed to prevent the spread of contaminants to clean areas of the site.

8.3 Regulatory bodies involved in contaminated land

Various organisations (as listed below) will be involved in granting approval for the treatment and the redevelopment of contaminated sites.

- ☑ Local Authority, which has a statutory duty to inspect its land and identify any sites that are formally designated as contaminated.
- ☑ Local Authority, in granting planning permission, will approve the remediation strategy to ensure that the ground is suitable for use of the proposed development.
- ☑ Environment Agency (EA) for England, Natural Resources Wales (NRW) for Wales, the Northern Ireland Environment Agency (NIEA) for Northern Ireland and the Scottish Environment Protection Agency (SEPA) for Scotland, for issuing an environmental permit for the use of mobile treatment plant to treat contaminated soils or for the reuse of construction and demolition waste.
- ☑ Local Authority for issuing an environmental permit for crushing equipment and for granting an exemption for the crushing and screening of demolition materials.
- ☑ Relevant environment agency for regulating the disposal of waste under the duty of care.
- ☑ Relevant environment agency for regulating contaminated sites that are deemed to be special sites (for example, defence sites or radioactive sites).
- ☑ Local Authority environmental health officers for dealing with any complaints regarding dust that crosses the site boundary.
- ☑ Water companies and the relevant environment agency for the disposal of polluted water from contaminated sites.

Effective contact with each of these authorities is essential and must be established at an early point in the project.

Any contaminated site must be totally fenced off and adequate warning notices must be prominently posted, advising all members of the public that the site is dangerous and to refrain from entering.

For further information on waste management refer to Chapter E03 Waste management.

8.4 Contaminated Land Regulations

Part IIA of the Environmental Protection Act (EPA) provides the legal framework for dealing with contaminated land in the UK. It is implemented in each of the devolved administrations through the following regulations.

E 08

8.4.1 England

- ☑ Contaminated Land (England) Regulations.
- ☑ Radioactive Contaminated Land (Modification of Enactments) (England) Regulations, which extends controls on contaminated land to radioactive contaminated land.

8.4.2 Northern Ireland

- ☑ The Contaminated Land Regime, which is set out in Part 3 of the Waste and Contaminated Land (Northern Ireland) Order has been enacted but is not yet in force.

This regime is very similar to that provided in Part IIA of the Environmental Protection Act in England, Scotland and Wales. The District Councils will be the primary regulators for the regime. Initial activity will focus on the preparation of site inspection strategies, which are to be completed within 12 months of the regime being in place.

8.4.3 Scotland

- ☑ SSI 2007/178 and statutory guidance were brought into force in 2000 and are similar to the regulations that apply in England and Wales.
- ☑ Contaminated Land (Scotland) Regulations.
- ☑ Radioactive Contaminated Land (Scotland) Regulations.

Under these regulations Local Authorities have a duty to inspect their land, identify whether it is contaminated and decide whether the land should be designated as a special site because of the nature of the contamination. Special sites are regulated by the Environment Agencies. All Local Authorities are required to draw up a contaminated land strategy.

8.4.4 Wales

- ☑ Contaminated Land (Wales) Regulations.
- ☑ Radioactive Contaminated Land (Modification of Enactments) (Wales) Regulations, which extends controls on contaminated land to radioactive contaminated land.

Contaminated land

For the purpose of the regulations ***contaminated land*** is defined as any land that appears to be in such a condition, by reason of substances in, on or under the land that significant harm or significant pollution of controlled waters is being caused or there is a significant possibility of such harm or pollution being caused.

Once a site is identified as contaminated land under the regulations, the Local Authority (or relevant environment agency in the case of a special site) has a statutory duty to ensure that remediation takes place by an appropriate person. Following designation of land as being contaminated land the regulatory authority should serve a remediation notice on the appropriate person(s) specifying what needs to be done and by when. The appropriate person(s) will be those who knowingly permitted the pollution or, if these cannot be found, the responsibility will fall to the owner or occupier of the site.

There is a close relationship between the contaminated land regime under EPA Part IIA above and planning controls.

The National Planning Policy framework confirms that planning policies and decisions should ensure the following.

- The site is suitable for its new use, taking account of ground conditions and land instability, including from natural hazards or former activities (such as mining), pollution arising from previous uses and any proposals for mitigation, including land remediation or impacts on the natural environment arising from that remediation.
- After remediation, as a minimum, land should not be capable of being determined as contaminated land under Part IIA of the Environmental Protection Act.
- Adequate site investigation information, prepared by a competent person, is presented.

Planning conditions should ensure appropriate investigation, remediation, monitoring and record keeping. Where contamination is suspected the developer is responsible both for investigating the land to determine what remedial measures are necessary to ensure its safety and suitability and for the actual remediation. There is a significant emphasis on voluntary remediation by the developer to avoid a formal remediation notice being issued where it meets the contaminated land definition.

Where there is a requirement to treat or dispose of contaminated material, then waste controls and the duty of care are likely to apply *(refer to 8.6 Managing contaminated land)*.

E
08

8.5 Brownfield sites

The definition of a brownfield site is very wide but generally relates to land that has had some form of previous development.

Many brownfield sites will be land that is affected by contamination but not to an extent that it automatically falls within the definition set out in the Contaminated Land Regulations above. Previously published estimates of the extent of land affected by contamination vary widely, from 50,000 to 300,000 hectares, amounting to as many as 100,000 sites. The Environment Agency estimates that, of these, 5,000 to 20,000 may be expected to be problem sites that require action to ensure that unacceptable risks to human health and the environment are minimised.

Sites with previous industrial occupation should be assumed to be polluted

Some brownfield sites are affected by land contamination because of the previous industrial uses of the site, which has led to the deliberate or accidental release of chemicals onto the land. This will depend on the type of activity previously being carried out. The following are examples of chemicals associated with four industrial processes.

- **Oil refineries** (fuel oil, lubricants, bitumen, alcohols, organic acids, polychlorinated biphenyls (PCBs), cyanides, sulphur and vanadium).
- **Lead works** (lead, arsenic, cadmium, sulphides, sulphates, chlorides, sulphuric acid and sodium hydroxide).
- **Pesticide manufacturing** (dichloromethane, fluorobenzene, acetone, methanol, benzene, arsenic, copper sulphate and thallium).
- **Textile and dye works** (aluminium, cadmium, mercury, bromides, fluorides, ammonium salts, trichloroethene and polyvinyl chloride).

8.6 Managing contaminated land

Managing land affected by contamination involves the identification of risks and then putting in the appropriate control measures to reduce those risks to an acceptable level so that the land is suitable for its intended use.

The process for dealing with contaminated sites should be in accordance with the following three stages.

Stage 1	Assessment to establish whether there are any unacceptable risks and, if so, what further action needs to be taken
Stage 2	Reviewing the remediation options and determining the most appropriate remediation strategy
Stage 3	Implementation of the preferred remediation strategy

These three steps are based on *Model procedures for the management of land contamination* (Contaminated Land Report 11 (CLR 11)), published by DEFRA and the Environment Agency, and are covered in more detail on the following pages.

8.6.1 Stage 1. Assessment of the risks from land contamination

The risks caused by land contamination are based on the concept of pollutant linkage, which is the relationship between the source of the contamination, the pathway that the contaminant could follow and the receptor. Each of these elements can exist independently but they have to be linked in order for there to be a risk.

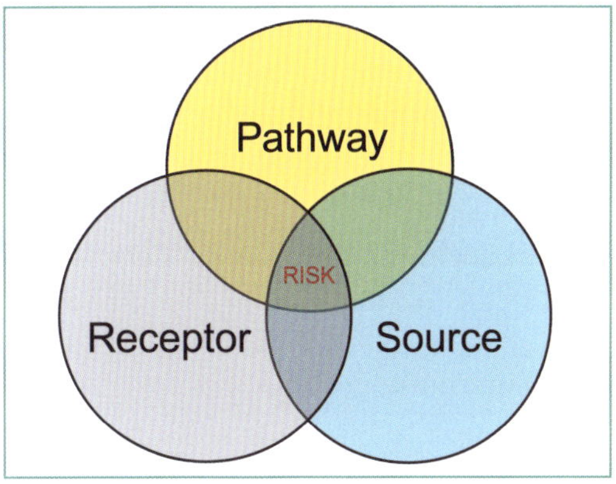

Every site will have different risks due to the type of contamination, geology and receptors, and acceptable levels will be dependant on the proposed end use for the site. The assessment may require a specialist to identify the level of risk.

Receptors can be any of the following.

- ☑ Humans (not just risks to site personnel but to end users following development).
- ☑ Controlled waters (surface waters and groundwater).
- ☑ Existing and potential ecosystems (plants and wildlife living in the site or close by).
- ☑ Property (building structures and services).

Examples of a source (contaminant), pathway and receptor are shown below.

- ☑ Source (oil, concrete washout and solvents).
- ☑ Pathway (groundwater, sewers and permeable land).
- ☑ Receptor (aquifer, animals and humans).

Establishment of the risks associated with the site will be based on information gathered from a preliminary (Phase I) investigation, which may include, but not be limited to, the following.

- ☑ History of the site.
- ☑ Previous processes, including their location, raw materials, products, waste residues, and methods of disposal.
- ☑ Layout of the site, above and below ground.
- ☑ Presence of waste disposal tips, made ground, abandoned pits and quarries with or without standing water.
- ☑ Mining history.
- ☑ Information on geology and hydrology.
- ☑ Potential uses of sites, past or present, in the area adjacent to the site.

This information is derived mainly from a desktop exercise, together with site reconnaissance as appropriate. This will allow a targeted site investigation (Phase II) to be undertaken if this is deemed necessary to determine the type, concentration and extent of any specific contamination.

Refer to BS 10175, *Investigation of potentially contaminated sites – Code of Practice*. The relevant recommendations and guidance within this standard are intended to ensure that the objectives of an investigation are achieved and that appropriate data for the risk assessment is obtained.

The Environment Agency has established its monitoring certification scheme (MCERTS) performance standard to deliver high quality environmental measurements. For chemical testing of soils where results are to be submitted to the Environment Agency the performance standard is an application of ISO/IEC 17025.

For a contaminated land checklist refer to Appendix A.

8.6.2 Stage 2. Appraisal of remediation options

The risk assessment described above will identify any unacceptable risks that will need to be reduced through the removal of existing pollutant linkages to make the site suitable for its new intended use. Remediation options will need to consider whether risks can be reduced at the source, pathway or receptor.

In-situ bioremediation treatment of contaminated land by introducing micro-organisms that remove pollutants by using them as food

- ☑ Source reduction – reducing, removing or breaking down the contaminant (for example, bioremediation).
- ☑ Pathway management – preventing the migration of contaminants to receptors (for example, installing a physical barrier or encapsulation).
- ☑ Exposure management – protecting the receptor (for example, by limiting the use of the land, such as preventing the growing of vegetables for human consumption).

Contamination identified through site investigation does not necessarily pose a threat if left undisturbed, and construction works should be planned to avoid disturbance if it has been decided to leave the material in situ. The best method for any given site will depend on a number of factors, which could include the following.

- ☑ Effectiveness (time taken and achievement of the standard of remediation).
- ☑ Cost of remediation.
- ☑ Practicality (technical, site, time and regulatory constraints and interaction with other works).
- ☑ Durability (the period of time that the remediation will need to be maintained).

It is now recognised that the previous practice of digging and dumping contaminated material by disposal to landfill is unsustainable. This is because the cost of landfill tax is rising sharply. Furthermore, tax relief associated with the off-site disposal of cleaning up contaminated land has been removed. It is also necessary to demonstrate that materials have been subject to treatment before being sent to a landfill site. There are strict limits on the levels of contaminants for material being sent to landfill (waste acceptance criteria).

For further information refer to Chapter E03 Waste management.

Treatment of contaminated materials on the site in which they occur is, in many cases, a cheaper option. The recovered materials can also provide a valuable resource, reducing the need to import clean virgin aggregates. There are often opportunities to reuse material after treatment in accordance with the waste hierarchy.

- ☑ Prevent waste (for example, by adjusting site layout to minimise waste quantities).
- ☑ Reuse (for example, on site reuse in an appropriate way).
- ☑ Recycle (for example, treat it to make material suitable for use on or off site).
- ☑ Dispose (for example, landfill or incineration).

Whatever remediation strategy has been identified there are a number of treatment options available. These are identified in *Model procedures for the management of land contamination* (CLR 11) and include the techniques shown in the diagram (contaminated land treatment options).

Three main types of remediation are containment, separation and destruction.

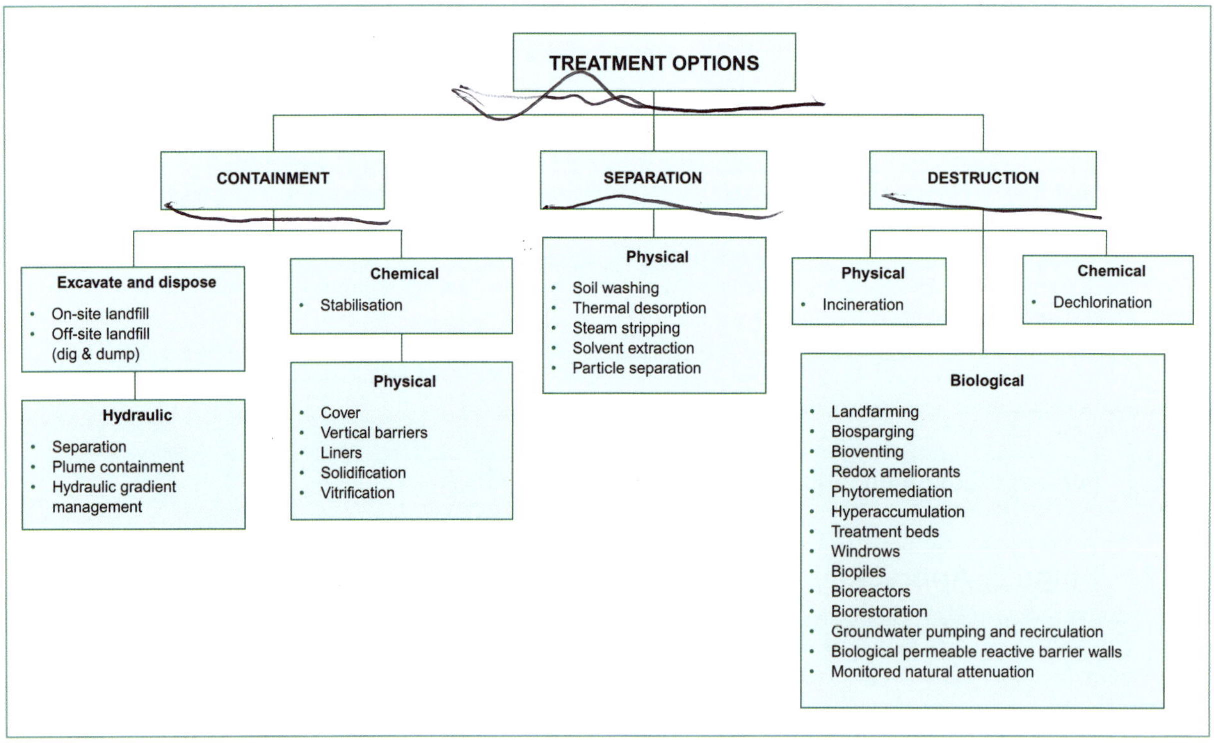

Contaminated land treatment options

8.6.3 Stage 3. Implementation of remediation options

The implementation of the preferred remediation option(s) will require consideration of the required permits and licences and discussions with the Environment Agencies and Local Authorities should take place as early as possible.

Many site remediation strategies will also include a gas or low permeability membrane system in the design to act as a final fail safe measure to protect the building and its occupants from potential future migrating contamination entering the building. This physical barrier will prevent such elements as methane (CH_4), carbon dioxide (CO_2), hydrocarbons, volatile organic compounds (VOCs) and radon (RN), which have all been identified to be hazardous to human health.

With increased development on brownfield sites, gas membranes are becoming a more common part of the build process. Understanding the membrane's importance is critical to all occupations that encounter it. The prevention of damage and introduction of working systems that prevent damage and/or protect the membrane should be considered in the build programme.

Gas membranes being installed

A standard notice used for membrane identification and issued as part of an induction

Strict planning conditions, along with British Standard guidance (BS 8485), will require verification of these membranes to ensure their integrity, which will have a major impact on the final condition being discharged and the building being occupied. Failing to protect it will cause timely delays and will be costly.

Early identification to all trades who directly encounter the membrane must be covered during the induction. The use of warning notices, delivery of toolbox talks and restricting access to completed areas will assist in the integrity of the membrane system and provide a safe environment for the occupants who will use the completed building.

For further information refer to the British Geomembrane Association website.

8.6.3.1 Waste controls for the treatment of contaminated land

Contaminated material that is excavated, recovered by treatment or disposed of may be classified as waste. Where it is waste then a range of waste regulations will apply.

The treatment of contaminated soil and/or contaminated waters may require a **mobile treatment permit** (England and Wales) or licence (Scotland and Northern Ireland). These permits are issued by the environment agencies.

A mobile treatment permit is used to regulate mobile plant equipment that involves treatment either in situ or ex situ. The permit identifies and sets out the type and extent of work that can be carried out. A site-based permit has to be used where a mobile plant permit is not applicable. The environmental permit can be either a standard rules permit or a bespoke one, depending upon the type of treatment and site location.

Operators who want to treat contaminated soil and/or contaminated waters using their mobile plant permit (MPP) at a particular site must also submit a site-specific deployment application (the deployment form and supporting information). The deployment application provides site-specific information and details the potential impacts arising from the proposed use of the mobile plant. The operator must demonstrate that the activity will not cause pollution of the environment, harm to human health or serious detriment to local amenities.

Following treatment under a mobile treatment permit or licence the material would normally cease to be waste providing it was excavated and treated on the site where it was used or is part of a remediation cluster. Where this is not the case then an environmental permit, licence or registered exemption would be required.

The CL:AIRE *Definition of waste: Development industry Code of Practice* (DiCOP) sets out good practice in England for dealing with contaminated land and defining when material ceases to be waste. It requires the implementation of a materials management plan (MMP) and a qualified person will declare that the Code of Practice (CoP) has been complied with.

A verification report at the end of the development will show that the material has been properly and suitably used and causes no harm to human health and the environment.

8.6.3.2 Removal of contaminated waste

Where materials cannot be incorporated into the site then the final option is disposal and all duty of care requirements must be met. Before the material can be removed it has to be properly classified to determine whether it is hazardous or non-hazardous waste. It will be necessary to demonstrate that treatment has taken place to reduce the quantity or physical nature of the material, through segregation or sorting.

Where contaminated waste and other materials are to be removed from a site, protective sheeting for skips and lorries will be necessary.

All skips or vehicles must be completely sheeted within the dirty area of a site. Care must be exercised by those carrying out the sheeting operations to ensure that they do not come into contact in any way with contaminated materials.

Vehicle drivers should not sheet their own vehicles, except to finally tighten sheet ropes, which should only be done in the clean area of the site.

Facilities must be available to thoroughly wash all vehicles leaving a contaminated site.

Detailed records must be kept of the disposal of hazardous or contaminated waste. Details should also be recorded within the project SWMP, as appropriate.

For further information relating to waste management refer to Chapter E03 Waste management, in particular Appendix D, which deals with reuse options for demolition materials and treatment of excavated soils.

8.6.3.3 Reducing pollution potential

Stockpiling of contaminated material should be avoided. However, where this has to be carried out the following precautions must be taken.

- ☑ The contaminated material should be placed on an impermeable surface or sheeting to avoid cross-contamination.
- ☑ Stockpiles should be covered to avoid dust and wind-blown contaminants.
- ☑ Access to stockpile areas should be restricted to authorised personnel.
- ☑ Stockpiles should be placed well away from drainage systems and watercourses.
- ☑ Silt fences may be required to prevent any run-off.

Exposure of contaminated materials in the ground should also be avoided but, where this is not possible, plan management work in the best season or weather to avoid the spread of contaminated dust or water.

Dewatering from site excavations on contaminated sites should also be considered carefully.

- ☑ Dewatering from excavations can draw water from contaminated adjacent sources.
- ☑ Discharges could be contaminated and must be disposed of with the consent of the appropriate authority.

Whilst proper site investigation will significantly reduce the likelihood of discovering contamination, if it is discovered unexpectedly there should be controls in place to deal with it. Obvious signs of contamination are listed below.

- ☑ Soil discolouration from chemical residues.
- ☑ Odours.
- ☑ Fibrous materials (such as asbestos).
- ☑ Chemical containers or tanks.
- ☑ Previous waste deposits.

Where contamination is suspected the following actions are required.

- ☑ Stop the works immediately.
- ☑ Report details of the discovery to site management.
- ☑ Prevent access to the area.
- ☑ Clear any fuels or substances in the vicinity that could cause fire or explosion.
- ☑ Contact the Local Authority or relevant environment agency when preliminary details of the contamination is known.
- ☑ Test the contamination to determine its exact nature and extent.
- ☑ Agree the appropriate remediation strategy with the Local Authority or relevant environment agency.

8.7 Occupational health considerations

The health of workers on contaminated sites can be affected through one or more of the following ways.

- ☑ Asphyxiation.
- ☑ Gasing.
- ☑ Ingestion.
- ☑ Inhalation.
- ☑ Skin absorption.
- ☑ Skin penetration.

Personal protective clothing (PPE) and approved respiratory protective equipment (RPE) must be worn at all times when work is carried out on contaminated sites.

Continuous assessments of the risk to health by exposure to any contaminated material or land must be carried out, and the control measures or precautions constantly monitored.

8.7.1 Personal hygiene

The level of risk to health by any contaminants will determine the scale of the need for hygiene facilities, but certain consideration should always be borne in mind when working on a contaminated site.

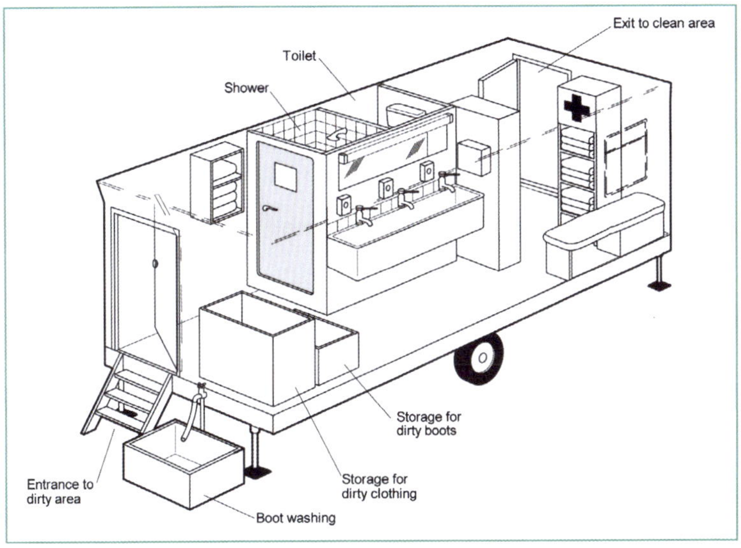

Typical layout of a hygiene unit, divided into three areas, with the dirty entrance remote from the clean exit

A dirty area is required for workers to discard dirty or contaminated clothing. Such clothing should be bagged and identified within this area before being dispatched to specialist cleaners.

Washing and toilet areas. Toilets, showers and washing facilities should be positioned between the dirty and clean areas, so that workers may wash or shower in order to remove any contaminant from their bodies.

A clean area is required for workers to put on clean and non-contaminated clothing. Access to and exit from this area must only be to the clean part of the site. It is essential that the entry and exit point of the clean area is in the clean part of the site.

Daily cleaning of the toilet facilities and the decontamination of all facilities must be carried out.

8.7.2 Asbestos

Asbestos was widely used in construction and insulation materials and poor management and waste practices have led to its contamination of brownfield sites. In addition, many unprotected sites have suffered from the indiscriminate fly-tipping of asbestos waste.

Asbestos is a highly dangerous material and presents substantial risks to the health of those who work with it and those who may be affected by it. Asbestos is a Category 1 carcinogen, responsible for around 8,000 deaths in the UK every year. Works dealing with asbestos come within the requirements of the Control of Asbestos Regulations.

Before the commencement of works, the site assessment should identify any risks posed by asbestos and the appropriate controls for dealing with it, which may range from physical containment to disposal.

For further information on the management of asbestos waste refer to Chapter E03.

The *Asbestos in soil and construction and demolition materials* joint industry working group (Asbestos in soil JIWG) was established in November 2011, after the Environmental Industries Commission (EIC) and CL:AIRE formally joined forces.

The aims of the joint industry working group are listed below.

- Bring together the asbestos management, occupational hygiene and brownfield management sectors with the aim of promoting the development of a consistent and harmonised approach to the regulation, investigation, analysis, assessment and management of asbestos in soil.
- Develop practical practitioner guidance on asbestos in soil that provides a consistent approach for UK industry, stakeholders and regulators.
- Promote the development of the relevant industry professional qualification framework for asbestos in soils for the brownfield and asbestos management sectors, building on the existing professional qualification framework for the management of asbestos in buildings, and relevant statutory requirements.
- Engage with the principal regulatory bodies for asbestos (the Health and Safety Executive (HSE), Environment Agency and representatives of Local Authority contaminated land officers) with the aim of promoting a consistent, unified and transparent regime for the regulation of all aspects relating to the remediation of land contaminated by asbestos.
- Promote and develop an improvement in public and stakeholder awareness of relevant issues, including health and safety, public health, technical, legal and insurance, related to the occurrence of and investigation and remediation of asbestos in soil.
- Promote the work of the Asbestos in soil JIWG to ensure all organisations are fully informed of its activities.

www.

For further information about managing and working with asbestos in soil and construction and demolition materials refer to the CL:AIRE website.

8.7.3 Explosives

Extreme care must be taken on sites where explosives are known to have been stored or used. This includes old mine workings, coal storage depots, former explosives factories and Ministry of Defence establishments. Disturbing any explosives could have sudden and disastrous consequences, especially if they are old and starting to decay.

Furthermore, unexploded bombs are occasionally unearthed when construction work takes place in areas that were subjected to bombing during World War II.

Once it is agreed that excavation work should proceed, this should be done with utmost caution. Any areas of soil discolouration, unusual objects or unusual cable presence should be taken as an indication that explosives are present. Work should be stopped immediately and the police informed.

Well established procedures already exist for competent military personnel to deal with unexploded devices.

8.7.4 Anthrax

Anthrax spores may lie dormant within soil, or in horse or cow hair binders in old lath and plaster, for many decades.

When such spores are disturbed, they still have the capacity to cause severe environmental problems. You should regard all premises (such as old tanneries, wool sorting stations and premises used in connection with animal carcasses, hides, bones, offal or for the production of gelatine, or old lath and plaster walls or ceilings) as high risk areas where anthrax spores may be present.

The Department for the Environment, Food and Rural Affairs (DEFRA) will be able to supply advice as to whether contaminated carcasses have been buried on old farm sites.

Where it is suspected that anthrax spores may be present, it is essential for everyone on the site to exercise good personal hygiene and use impervious personal protective clothing, including gloves. Any cuts and scratches that occur before work starts must be adequately covered. As an additional safeguard, advice on immunisation and general health procedures should be sought from a doctor.

8.7.5 Radiation

Before starting work on a site where any work involving radioactive materials has taken place previously, and where radioactive contamination (whether natural or artificial) may be present, consult the HSE and the relevant environment agency.

It will be necessary to use specialist contractors for all aspects of both the removal of substances and decontamination where radioactive materials are being dealt with.

Appendix A – Contaminated land checklist

Company name		Project title	
Location		Contract no.	

Contaminated land	Yes	No	N/A
1. Has a preliminary or desktop investigation been carried out of the site and its previous uses to identify the risks from contamination?			
2. Has a site investigation been carried out to identify the type and extent of any suspected contamination?			
3. Are the testing company and laboratory suitably competent to carry out the testing and analysis of any contamination?			
4. Has the proposed remediation strategy been agreed by the local planning authority and relevant environment agency, where appropriate?			
5. Does the site induction cover details of contamination and the procedures for working in areas of known contamination?			
6. Are there procedures in place to ensure that work is stopped and appropriate reporting takes place if contamination is discovered or accidently disturbed?			
7. Are areas of contamination fenced off to prevent vehicles and plant spreading the contamination across the site?			
8. Are relevant environmental permits in place for the treatment and remediation of contamination?			
9. Are suitable controls in place for dealing with contaminated water (such as discharge consents to foul sewers or tanker)?			
10. If contaminated materials have to be stockpiled, are they covered or provided with silt fences to prevent run-off and wind-blown contamination?			
11. Are contaminated materials stored on hard-stand areas to avoid contamination of ground and groundwater below?			
12. Are stockpiles positioned well away from drainage systems or watercourses?			
13. Are there suitable washing facilities to clean contaminated vehicles before leaving site (not just wheel-washing)?			
14. Where vehicles remove contaminated materials, are they covered with sheets before leaving site?			

Comments

Name		Position		Signature		Date	

For an interactive PDF version of this form visit the companion website.

E 08

09

Archaeology and heritage

Contents

Supporting information

G: Checklists and forms

Overview

The heritage environment is a precious resource bringing millions of tourists to the UK to enjoy our rich, cultural past. Archaeology is a major factor in construction and development and an estimated £150m is spent by developers on archaeology in the UK each year.

This chapter gives a general overview of the legal framework for the protection and management of historic buildings, monuments, archaeological sites and remains. It identifies which types of feature are protected in law and where to obtain consent to carry out works on or near them.

The chapter also gives some guidance on the necessary actions to be taken to protect historic and archaeological features during the works and if an unexpected archaeological discovery is made.

9.1 Introduction

Archaeology is often underestimated as a business or project risk so is often not considered early enough in the feasibility or design stages of projects, leading to unexpected or unplanned consequences.

Archaeological remains and the built environment provide a valuable record of a nation's history and identity and are an irreplaceable part of its national heritage. For these reasons archaeology and built heritage form an important element of planning policy and must be considered early in any construction project. Today, archaeology is a significant element of construction and development, and all parties (developers, archaeologists, conservation officers and the regulatory authorities) should aim to ensure that good practice is applied.

The main reasons why the construction sector should address archaeology are shown below.

- ☑ **Planning law, heritage law and planning policy.** Archaeological remains are an important part of our cultural heritage, and they are a fragile, finite and irreplaceable resource that needs to be protected. This is recognised within the UK planning process and by national legislation and guidance. Whilst some archaeological sites are protected by law, all archaeological remains affected by development under planning are treated as material considerations in the planning process.
- ☑ **Education and new knowledge.** Archaeology makes a significant contribution to education, social cohesion and the economy and it brings wide public benefits and knowledge through archaeological work in schools and universities.
- ☑ **Economy and society.** Archaeology underpins the UK tourism and heritage industries and creates jobs. It also plays an important role in regeneration by invoking a sense of place, belonging and cultural identity for new developments.
- ☑ **Sustainability.** Developers can demonstrate their commitment to sustainability and responsible development through a proactive approach to archaeology, and also a recognition that archaeological remains are a non-renewable resource that need environmental protection and management.

E 09

An early, positive and proactive approach to archaeological remains can:

- ☑ minimise the impacts
- ☑ provide early effective risk management in the planning phase, minimising the risk of unexpected archaeological finds, which would have significantly higher costs attached if found during the construction phase
- ☑ heighten a sense of place for new developments
- ☑ provide opportunities for involving the community, leading to positive publicity for the construction project.

9.2 Important points

An archaeological statement (as part of an overall environmental statement) may be required to support a planning application and may also require some intrusive works on the site. This is likely to include an archaeological management plan to identify the controls that are required during the works.

A professional archaeologist may be required, as a condition of any planning consent, to supervise the works as they proceed and to ensure that suitable protection measures are put in place and maintained.

An archaeologist may also be required to excavate and record remains that are due to be removed.

Suitable controls to manage archaeology and heritage should be included in the project as early as possible for the following reasons.

- ☑ To comply with legal requirements relating to scheduled monuments, listed buildings and protection of the historic environment.
- ☑ To ensure buildings are designed to avoid the disturbance of remains and to preserve historic features.
- ☑ To enable designers to incorporate historic features of the site in the final development.
- ☑ To avoid disturbance during the construction process itself by the following means.
  - – Ensuring all archaeological works are segregated from the main works with authorised entry.
  - – Identifying existing underground services that might have an impact on archaeology.
  - – Ensuring archaeological excavations are suitably protected during the works.
  - – Avoiding dewatering work in the vicinity of archaeological remains.

9.3 Protected monuments, buildings and sites

There are a series of designations and/or statutory protection measures for archaeological and historic remains that are closely linked with planning regulations. In March 2012 the Government published the *National planning policy framework* (NPPF), which superseded *Planning for the historic environment in England* (PPS5). On 1 April 2015 English Heritage was divided into two parts: Historic England, who inherited the statutory and protection functions of the old organisation, and the new English Heritage Trust, a charity that operates the historic properties and has taken on the English Heritage operating name and logo. Historic England has published *Good practice advice in planning* (Note 3). This replaces PPS5, which has now been withdrawn.

Similar planning controls will apply in Scotland, Wales and Northern Ireland.

For further guidance on planning in England, Northern Ireland, Scotland and Wales visit the relevant website.

9.3.1 Scheduled monuments

The word *monument* covers a whole range of archaeological sites. Scheduled monuments are not always ancient, or visible above ground.

Nationally important sites, buildings and monuments are given legal protection by being placed on a schedule for monuments. The UK heritage bodies identify sites that should be placed on the schedule by the respective government. The Ancient Monuments and Archaeological Areas Act supports a formal system of scheduled monument consents for any works to a designated monument.

All work to a monument that is scheduled as an ancient monument requires scheduled monument consent, as do any works to the grounds that surround it.

The following are criminal offences, potentially leading to fines or imprisonment.

- Carrying out unauthorised works on a scheduled monument without consent.
- Causing damage to a scheduled monument.
- Failing to adhere to the terms of a consent.

Development or construction works that affect a scheduled monument require formal permission (scheduled monument consent).

The granting of scheduled monument consent does not imply planning permission, or vice versa.

Stonehenge

Applications for scheduled monument consent are made:

- in England, to the Department of Culture, Media and Sport, through Historic England
- in Northern Ireland, to the Department for Communities
- in Scotland, to Historic Environment Scotland
- in Wales, to Cadw.

9.3.2 Listed buildings

A listed building has statutory protection against unauthorised demolition, alteration and extension.

In the UK there are three grades of listed building. For England and Wales the grades are shown below.

- **Grade I** are buildings of exceptional interest.
- **Grade II*** are particularly important buildings, of more than special interest.
- **Grade II** are buildings of special interest, justifying every effort to preserve them.

In Scotland the above grades are designated as A, B and C. In Northern Ireland these grades are designated as A, B+ and B. In both cases 'A' is the highest level of protection matching with Grade I in England and Wales.

Any works to be undertaken to a listed building that would affect its character, or the character of any building or structure in its curtilage (area of land attached to a building), require a listed building consent. It is a criminal offence to demolish, alter or extend a listed building without consent. If this occurs and the character and appearance of the building is affected, a listed building enforcement notice may be served, requiring reinstatement of the building prior to works being carried out. If this is not feasible a fine, sentence or both may be served.

It should be noted that Planning Authorities and National Park Authorities have the power to serve a building preservation notice on the owner of a building that is not listed, but which they consider is of special architectural or historic interest. The building is then protected for six months, like a listed building, in which time the Secretary of State must decide whether to list the building.

E 09

9.3.3 Conservation areas

A conservation area has a special architectural or historic character that is desirable to preserve or enhance and is designated as such by the local Planning Authority. Conservation area designation and policy seek to preserve or enhance the character of the whole area, including open spaces, not just individual buildings.

Conservation areas are given greater protection by the Local Authority, which has extra control over demolition, minor developments and the protection of trees. A conservation area consent from the local Planning Authority must be obtained prior to works being carried out in the designated area. The local Planning Authority should be consulted to determine if works to be carried out require a conservation area consent.

9.3.4 Parks and gardens

Historic parks and gardens can be registered, which means inclusion on the register of parks and gardens of special historic interest. Alterations to parks and gardens generally do not require statutory consent unless a planning application is required or it affects a tree covered by a tree preservation order. It should be noted that local Planning Authorities are encouraged to include policies in their development plans for the protection of registered historic parks and gardens.

9.3.5 Designated wrecks

Section 1 of the Protection of Wrecks Act provides protection to some 60 wrecks around the UK, designated for their archaeological and historical significance. In Northern Ireland these are regulated under the Historic Monuments and Archaeological Objects (Northern Ireland) Order.

The administration of this Act and the issue of licences allowing diving, survey, collection of objects or excavation on these sites is the responsibility of the following organisations.

- ☑ Historic England in England.
- ☑ Historic Environment Scotland in Scotland.
- ☑ Department for Communities in Northern Ireland.
- ☑ Cadw in Wales.

9.3.6 Designated vessels and controlled sites

Certain military aircraft crash sites and military shipwrecks are designated and protected under the Protection of Military Remains Act. These sites are administered by the Ministry of Defence.

9.3.7 Burial grounds and human remains

In England and Wales, authorisation to disturb human remains may be by Act of Parliament, Home Office licence (through the Ministry of Justice), or Church of England Faculty. Under Scottish law there is no such system of authorisation. In England, disturbance of human remains may be allowed by an Act of Parliament that authorises a specific major project, but otherwise the following legislation applies.

- ☑ Town and Country Planning (Churches, Places of Religious Worship and Burial Grounds) Regulations.
- ☑ Disused Burial Grounds Act.
- ☑ Disused Burial Grounds (Amendment) Act.
- ☑ Burial Act.

If any form of construction works is to take place on a disused burial ground then the Disused Burial Grounds (Amendment) Act stipulates that human remains should be removed before work begins. However, if planned works will leave human remains undisturbed then dispensation can be obtained from the Home Office authorising that the burials remain in situ.

9.3.8 Treasure

Under the Treasure Act (England, Wales and Northern Ireland), certain finds are deemed as treasure. All finds must be reported to the nearest police station or coroner for the district within 14 days, which will then advise on the course of action to be taken. Under the Act, the following finds would be classed as treasure.

Using a metal detector

- ☑ Coins that are at least 300 years old (coins are usually only treasure if there are ten or more, or if they are with other items that meet the treasure definition).
- ☑ Objects containing at least 10% of gold or silver and at least 300 years old.
- ☑ Any object that is found in the near vicinity of known treasure.
- ☑ Base metal deposits of prehistoric date.

In Scotland any treasure finds are controlled by an organisation called Treasure Trove in Scotland.

E 09

9.4 Managing archaeology

In risk terms, archaeological remains should be considered as hazards which, if not properly identified and mitigated, may cause an adverse effect on a development. Typical archaeological risks faced by development include:

- ☑ planning constraints and potential for spot listing and scheduling of remains
- ☑ unforeseen or unexpected archaeological remains affecting design, implementation programme and profit margins.

The environmental statement and planning conditions for the project, provided by the client, should identify any obligations for the management of archaeology and heritage. These requirements should be incorporated into the project environmental management plan (EMP). In some cases it will be a condition of the planning consent to prepare an archaeological management plan, which may include the employment of a qualified archaeologist as a watching brief during relevant construction works, to identify:

- ☑ what planning policy and legislation authorities will consider when deciding how to treat any archaeological remains
- ☑ the types of archaeological site and range of features that may occur
- ☑ the range of non-intrusive and intrusive techniques that archaeologists will use to assess and evaluate archaeological remains
- ☑ the main roles of the client, archaeological consultant, contractor and curator.

It is often best to deal with archaeological issues (including excavation) in advance of the main construction phase.

All field evaluation work should be completed and an agreed mitigation strategy (if required) should be in place before construction or groundwork starts.

It will also be important to agree the appropriate method of working adjacent to sensitive areas as vibration from operations (such as excavation or tunnelling) may cause damage. There may be an obligation to provide vibration monitoring of the works to ensure that vibration levels are not exceeded.

All areas of known archaeological or historic interest should be protected with suitable fencing to prevent damage or encroachment, with access to other site traffic and personnel carefully controlled. This is a legal requirement for scheduled sites.

In respect to health and safety, any excavations for archaeological works should be treated as normal and protected to avoid inadvertent access by personnel or site equipment. The location of underground services should also be identified as these could cross sites where archaeological excavation is planned.

Many archaeological excavations will also take place on brownfield sites that have been contaminated. If there is reason to believe that the ground is contaminated, arrangements should be made to undertake sampling and testing before archaeological work starts on site.

Dewatering schemes can have a negative impact on archaeological features as they could cause differential settlement or damage to materials that have previously been protected by being waterlogged. Appropriate methods of dewatering should be agreed in advance of the works taking place.

9.5 Unexpected discovery

If suspected archaeological objects or remains are found during the construction works without an archaeologist on site, certain procedures should be followed. Archaeological finds should not be disturbed any further without any specialist investigation and advice. Advice should be sought by contacting the Local Authority archaeological officer on how to proceed.

Accidental discovery

In the case of accidental discovery of archaeological finds or human remains the following actions are considered good practice.

- ☑ Stop work in the area of the discovery.
- ☑ Leave the find in situ and undisturbed.
- ☑ Control access to the area to authorised persons only.
- ☑ Stop vehicle traffic entering the area.
- ☑ Report the find to the site manager.
- ☑ Take specialist advice as appropriate.
- ☑ Report human remains, treasure and other archaeological finds to the appropriate statutory authority (usually the police or the coroner).
- ☑ Report the find to the appropriate local planning authority or other curator.

Under the Burials Act there is a requirement to report unexpected discoveries of human remains to a coroner. If there is any possibility that the remains are recent (for example, less than 100 years old) the local police should be contacted to determine whether the remains are ancient or not. For authorisation to continue, approval will need to be given by the Home Office.

For an archaeology and heritage checklist refer to Appendix A.

Appendix A – Archaeology and heritage checklist

Company name		Project title	
Location		Contract no.	

Archaeology and heritage	Yes	No	N/A
1. Has the contract documentation been reviewed to determine the presence or location of archaeology or protected heritage and buildings?			
2. Has an archaeological investigation been carried out?			
3. Have designers been involved to protect or integrate known archaeological features into the project?			
4. Have licences or permissions (such as scheduled monument consent or listed building consent) been obtained to carry out work on or around archaeology or protected heritage and buildings?			
5. Where there are works on or around known archaeology or heritage, has an archaeologist been employed to carry out a watching brief to monitor and record the works?			
6. Have locations containing historic sites or archaeology been adequately fenced off or protected?			
7. Have suitable communication arrangements been put in place for unknown finds that may be discovered, and for reporting these to the local archaeological officer or Local Authority, and recorded?			
8. Have photographs and records been taken of the existing historic elements prior to work commencing on site?			
9. Has the information relating to archaeology and historic elements been provided in the induction?			
10. Do operatives hold the appropriate qualification (such as a CSCS heritage card) and certificate of competency to work on historic buildings?			

Comments

Name		Position		Signature		Date	

For an interactive PDF version of this form visit the companion website.

E 09

10

Site environment management systems

Contents

Supporting information

G: Checklists and forms

Overview

This chapter provides a general introduction to environmental management systems, how they can be applied to construction projects and guidance on the nature and type of environmental documents and records that should be maintained on a construction site.

10.1 Introduction

An environmental management system (EMS) can help companies reduce their environmental impact, achieve cost savings, comply with legislation and demonstrate their commitment to securing continual improvement in environmental performance.

Being able to demonstrate that their company understands and is managing its environmental impact is becoming an important prerequisite for many construction companies when tendering for new business.

It is recognised that adopting standard environmental management policies and practices not only helps in protecting the environment but also brings business benefits in terms of reduction of waste, energy use and improved efficiency, all of which are aligned to the Green Construction Board's simple message of ***Reduce carbon = Reduce cost***. It also reduces the risk of causing incidents that may result in enforcement action, which could lead to prosecution.

Managing the environment, or environmental impact, is not just for one person in an organisation. It is a process of change for all involved.

An EMS can help all types and sizes of company meet their own environmental and sustainability targets as well as contribute to national targets on climate change, sustainable development, waste, water, emissions, energy, resource efficiency and other environmental issues.

An EMS has a number of important functions.

- ☑ **Help a company understand its environmental impact.** It should be a practical tool to identify and describe a company's impact on the environment, manage and reduce these impacts and evaluate and improve performance. An EMS can help with identifying compliance obligations and managing risks and opportunities.
- ☑ **Secure positive environmental outcomes.** It should not only document procedures and processes but also focus on improving environmental performance and complying with legal and client requirements.
- ☑ **Reduce costs and improve efficiency.** An EMS can also help conformity with customer requirements in the supply chain, enable sustainable procurement policies, enhance a company's reputation, secure new markets and help improve communication with employees, regulators, investors and other stakeholders.

To fully contribute to improved environmental performance, a good EMS should include the detail below.

- ☑ Be implemented at a senior level and integrated into company plans and policies (top-level leadership and commitment is required so that senior management understand their role in ensuring the success of an EMS).
- ☑ Identify the company's impact on the environment and set clear objectives and targets to improve its management of these aspects as well as the company's overall environmental performance.
- ☑ Be designed to deliver and manage the organisation's compliance with environmental laws and regulations on an ongoing basis. This will quickly instigate corrective and preventative action in cases of legal non-compliance.
- ☑ Deliver good resource management and financial benefits.
- ☑ Incorporate performance indicators that demonstrate the effectiveness of the above and can be communicated in a transparent way in annual reports.

Demonstrating to clients that you have policies and arrangements in place to manage and improve environmental performance will enhance your company's reputation and contribute to obtaining future work.

Whilst larger construction companies will want to demonstrate that they have a robust environmental management system through certification to a formal standard (such as ISO 14001 *(refer to 10.3)*), smaller companies should consider implementing an EMS as this will help identify their environmental risks and put in controls to manage them. This will also improve their opportunities in winning future work.

10.2 Important points

An effective environmental management system (EMS) helps companies to manage their environmental risks, reduce their exposure to potential legal action and improve performance. This leads to other benefits, such as cost and efficiency savings and improved relationships with customers.

An EMS can also help companies demonstrate that they have policies, objectives and programmes in place to meet their own sustainability objectives, as well as contributing towards wider Government targets.

E 10

Important aspects for delivering an effective EMS on site are shown below.

- Identifying the environmental and legal obligations for a project from a review of client documents. This is including, but is not limited to, planning consents and conditions, environmental statements, remediation statements, archaeological statements and ecological management plans.
- Identifying relevant environmental aspects (such as discharges to land, air and water) and meeting these obligations and their associated risks. This should also include considering emergency situations (such as leaks and spillages).
- Putting the appropriate responsibilities in place to manage and reduce the impact associated with the project environmental obligations and risks.
- Creating an EMS plan that sets out what actions need to be taken, who is responsible for those actions, and identifying the appropriate records that need to be maintained.
- Putting an appropriate inspection, monitoring and audit regime in place to ensure that the project's environmental obligations and their associated risks are being managed effectively.
- Maintaining and retaining appropriate records to demonstrate that legal, client and company requirements, associated with the project, are being met and can be retrieved, if required, for any purpose.

10.3 Types of environmental management system

There are three main recognised standards or schemes.

- **ISO 14001** is the international standard for EMS, which specifies the components necessary to help organisations to systematically identify, evaluate, manage and improve the environmental impacts of their work, products and services.
- **EMAS** (the EU Eco-management and audit scheme) is a voluntary EU-wide environmental registration scheme that requires organisations to produce a public statement about their performance against targets and objectives, and incorporates the international standard ISO 14001.
- **BS 8555** is a British Standard, published in 2003, which breaks down the implementation process for ISO 14001 or EMAS into six stages, making implementation much easier, especially for smaller companies. The Institute of Environmental Management and Assessment (IEMA) has developed the IEMA Acorn scheme, which enables companies to gain UKAS accredited recognition for their achievements at each stage of the standard as they work towards ISO 14001 or EMAS. This process allows early recognition of progress against indicators and can be used very effectively to enhance supply chain management by setting agreed levels of performance, certified to a national standard, and which have been checked by an independent auditor. Organisations who complete each stage of the scheme are entered on a public Acorn register.

ISO 14001 requires organisations to meet the following requirements.

- Identify interested parties relevant to the EMS and their needs and expectations.
- Ensure that top management demonstrate leadership and commitment with respect to the EMS.
- Establish an environmental policy relevant to the organisation.
- Identify the environmental aspects associated with the organisation's work and determine the significant environmental impacts.
- Identify applicable compliance obligations associated with its environmental impact.
- Identify priorities and set appropriate environmental objectives.
- Identify and provide the necessary resources for the establishment, implementation, maintenance and continual improvement of the EMS.
- Establish a structure and programme(s) to implement the policy and achieve environmental objectives.
- Implement planning, control, monitoring, preventive and corrective actions, auditing and reviewing procedures to ensure that the policy is complied with and that the environmental management system remains appropriate, and capable of adapting to changing circumstances.

ISO 14001:2015

ISO 14001:2015 was published on the 15 September 2015. The main changes are outlined below.

- Increased prominence of environmental management within the organisation's strategic planning processes.
- Greater focus on leadership.
- Addition of proactive initiatives to protect the environment from harm and degradation (such as sustainable resource use and climate change mitigation).
- Lifecycle thinking when considering environmental aspects.
- Addition of a communications strategy.

Accredited organisations are granted a three-year transition period after a revision has been published to migrate their environmental management system to the new edition of the standard. After this transition period, companies that opt for third-party certification will have to seek certification to the new version of the standard (ISO 14001:2015).

- For further information on ISO 14001:2015 refer to the ISO website.
- For further information and a guide to environmental management systems refer to the WRAP website.

For a site environment management systems checklist refer to Appendix B.

Other standards have been introduced to manage more specific environmental issues, such as energy management under ISO 50001. ISO 50001 provides a framework of requirements enabling organisations to meet the following.

- Develop a policy for more efficient use of energy.
- Fix targets and objectives to meet the policy.
- Use data to better understand and make decisions concerning energy use and consumption.
- Measure the results.
- Review the effectiveness of the policy.
- Continually improve energy management.

ISO 50001 certification can be used to comply with the energy savings opportunity scheme (ESOS), as this is sufficient to constitute an ESOS assessment.

For further details of the ESOS scheme, which is relevant to larger companies, refer to Chapter E04 Energy management.

10.4 Policy, objectives and targets

Requirements for a basic EMS policy are shown below.

- Be appropriate to the organisation, for its size and type of work.
- Establish a framework for setting and reviewing environmental objectives.
- Commit to comply with current legislation and other environmental requirements or obligations.
- Commit to pollution prevention and continual improvement.
- Be documented, implemented and maintained.
- Be readily available for employees and other interested parties.

The policy should be endorsed by someone of authority within the organisation, either a director or someone of equal seniority.

Setting objectives is vital to a successful EMS and demonstrates the commitment of the business to reduce its impacts and set the path for a programme of continual improvement.

An initial baseline assessment of where the organisation is in terms of its current management of the environment will identify areas for improvement.

The following areas could be considered.

- Measuring and reducing the amount of waste produced.
- Improving recycling rates for different waste streams.
- Monitoring water use, setting reduction targets and implementing and sharing measures.
- Monitoring energy use (electricity and fuel), setting reduction targets and implementing and sharing measures.
- Purchasing services and materials from sustainable sources and/or with recycled content.
- Assisting and improving supply chain knowledge on environmental matters.

Regular measurement and reporting against company objectives will highlight whether they are being met and where future action needs to be focused. This will allow the company to continually improve its environmental performance and demonstrate to clients, staff and the general public that it is taking its environmental responsibilities seriously.

For typical environmental objectives for a construction project refer to Appendix A.

10.5 Implementing environmental management on site

A company's environmental management system will define what plans and procedures, together with the appropriate forms, will need to be completed at a site level. Whatever documents are used to achieve this, there are some clearly defined steps in order to deliver effective environmental management on site. The following five steps should be followed.

10.5.1 Step 1. Identify the project environmental obligations

The first important step in managing the project environmental issues is to identify the environmental obligations. An obligation is a requirement to take some course of action, whether legal or moral. In the case of environmental obligations there are a number of potential sources.

- ☑ The first of these is legal obligations where individuals and organisations have to follow legal requirements or face prosecution (for example, duty of care under waste law or the protection of wildlife).
- ☑ The next concerns contractual obligations and these might relate to performance requirements that a client specifies (for example, BREEAM or CEEQUAL). Where these are not met a client can resort to a civil action to recover costs for the damages incurred.
- ☑ Another type of obligation arises from businesses' corporate responsibility and their duty to act in a way that recognises the interests and views of other stakeholders, in relation to the environment. Failure to recognise corporate responsibility can result in damage to corporate reputation.

A review of the project's contractual documentation, including any associated planning conditions or Section 106 agreements, should be undertaken to identify the project-specific environmental obligations. It is essential to have processes in place to identify potential obligations that might arise from different emergency scenarios that could have environmental impacts, and have an effective response plan in place that is regularly tested to mitigate effects if they occur.

10.5.2 Step 2. Identify the project environmental aspects and risks

Following the identification of the project environmental obligations the next step is to identify the associated environmental aspects and risks, together with any relevant opportunities. Environmental aspects and risks can be present on a specific project wherever there is an interaction with the environment, arising from either raw material inputs or emissions and other outputs to the environment. The goal of a good environmental management system is to identify what these aspects, risks and opportunities are and to put in place the necessary controls to manage them to within acceptable limits to achieve the desired objective.

- ☑ An **aspect** is usually associated with a 'doing word' (such as disposing of waste, discharging of dirty excavation water or storing fuel).
- ☑ The **risk** (or impact, as defined in ISO 14001) is the consequence of not doing something correctly.

e.g. Aspect and risk

Illegally disposing of waste is the aspect, and the associated risk is it being disposed of incorrectly, such as at an unlicensed tip.

Discharging water into a drain is the aspect; the risk is discharging water without consent or the discharged water polluting a connected river.

Another common aspect is storing fuel; the risk can be that fuel leaks into the ground and pollutes the ground or an underground drinking water aquifer.

The minimum level of performance is compliance with legal and other requirements. Organisations may also decide that they want to work to good practices where risks are minimised as far as reasonably practicable. The best way of managing environmental risks is in a systematic way and ISO 14001 *Environmental management systems* provides an internationally recognised framework for doing this.

The production of a project environmental aspects register, identifying the obligations, together with the associated risks and control measures (including emergency situations or possible worst case scenarios), will be an important tool for communicating these issues to contractors and site personnel.

Environmental risks and opportunities should be assessed during the pre-construction phase of a project to ensure that environmental management is properly integrated within the project with respect to these issues. Environmental risks are identified in the pre-construction phase through a number of sources, including ecology surveys, desk top studies and ground investigations. The client may also be in possession of existing environmental issues. All new surveys and existing information must be included in the pre-construction information and passed to the designers and contractors to enable them to carry out their duties.

10.5.3 Step 3. Identify the environmental responsibilities

Having defined the project environmental obligations and associated risks and opportunities it will be necessary to identify the main responsibilities for their control. An environmental management system can involve every person on a project or within an organisation. Certain people within the management system will have specific responsibilities and it is important that these are clearly defined and set out.

Examples of functions and positions with specific responsibilities are shown below.

- ☑ Directors.
- ☑ Environmental co-ordinators.
- ☑ Architects and designers.
- ☑ Noise specialists.
- ☑ Waste co-ordinators.
- ☑ Sub-contractors.
- ☑ Suppliers.
- ☑ Community liaison staff.

It is important to ensure that the necessary lines of communication are defined between those individuals that have prepared the project environmental plan and site personnel.

Responsibilities of an environmental manager on a large construction project

- ☑ Implementing the requirements of the company EMS.
- ☑ Ensuring that relevant environmental policies are displayed and communicated to all project staff and personnel.
- ☑ Ensuring that a project assessment is carried out to identify the main types of work and aspects.
- ☑ Ensuring that an environmental management plan (EMP) is developed and maintained to identify the specific environmental requirements and responsibilities and includes legal, client and other relevant issues.
- ☑ Ensuring that a waste management plan is developed and maintained to manage waste and includes appropriate responsibilities, waste targets and legal compliance information.
- ☑ Ensuring that, as well as waste targets, relevant environmental objectives are set, implemented and monitored in line with the project construction programme or establishment requirements.
- ☑ Ensuring that the general environmental requirements and objectives are included in inductions, toolbox talks and briefings, and records are maintained.
- ☑ Ensuring that specific activity method statements, including those of suppliers, are reviewed to include the relevant project environmental and waste requirements.
- ☑ Ensuring that appropriate inspections and monitoring arrangements are put in place to meet the environmental requirements (such as weekly supervisors' inspections).
- ☑ Ensuring that appropriate environmental emergency arrangements are identified, implemented and tested, as appropriate.
- ☑ Participating in management reviews and audits of the status, adequacy and effectiveness of the project EMP.
- ☑ Managing environmental non-conformances and subsequent corrective actions.
- ☑ Reviewing advised changes to environmental legislation and other requirements and taking the appropriate action.
- ☑ Ensuring that an environmental management file (EMF) is established to contain appropriate records of the above and is sufficient to meet the environmental requirements.

10.5.4 Step 4. Create an environmental management plan

A project EMP is a vital tool for setting out what actions need to be taken and who is responsible for them. It will contain information including method statements, legislation, performance requirements and environmental risks. The typical contents of a project EMP may include, but not be limited to the following.

- ☑ Environmental policies.
- ☑ Project environmental objectives and targets.
- ☑ Environmental appointments.
- ☑ Environmental risk assessments.
- ☑ Site waste management plan.
- ☑ Environmental emergency response/action plans.
- ☑ Consultation.
- ☑ Environmental inductions, training and awareness.
- ☑ Environmental monitoring, measuring, inspections and audits.
- ☑ Environmental incident and investigation reports.
- ☑ Project environmental records.

10.5.5 Step 5. Monitor and inspect

After implementing a project EMP a robust monitoring and inspection regime should be put in place to ensure that the identified risks and opportunities are being managed and that legal and contractual requirements are being met. Examples of monitoring and inspection work for risk management and compliance include the following.

- ☑ Noise measurements.
- ☑ Water analysis.
- ☑ Air quality.
- ☑ Dust monitoring.
- ☑ Duty of care checks.

Monitoring and inspection can also take place to assess performance delivery and the achievement of performance objectives that have been set by a project. This may include monitoring and inspection of energy, carbon, water, resource and material use efficiency.

Monitoring and inspection work can include the following.

- ☑ Director tours.
- ☑ Site management tours.
- ☑ Supervisor tours.
- ☑ Internal audits.
- ☑ External audits.
- ☑ Supplier audits.
- ☑ Water monitoring.
- ☑ Energy usage.
- ☑ Resource use measurement.

10.5.6 Environmental performance improvement

Environmental performance improvement is vital to successful construction projects. Efficient use of energy, carbon, water, raw materials and waste reduction will bring multiple benefits to projects and contribute to cost savings. To achieve performance improvement requires a well-planned and structured management system, the right leadership and a mindset across the project that results in behaviours that are focused towards resource efficiency.

Regular project reviews involving all site personnel that are identified in the project EMP, including sub-contractors and suppliers, will provide the opportunity to discuss environmental issues and how to improve performance. These reviews need not be separate, formal meetings and could be an agenda item of standard management meetings.

10.6 Environmental documentation

The project environmental requirements will dictate what environmental documentation is required to demonstrate compliance. It is likely that a company will have a standard project filing system and it would be good practice to ensure a section of this is reserved for environmental records.

Some likely documents that will need to be retained on site

- ☑ Initial review and identification of project environmental obligations.
- ☑ EMP, including schedule of aspects, risks and details of their control.
- ☑ Waste management plan.
- ☑ Details of emergency environmental arrangements, including drainage plans and location of manholes.
- ☑ Copies of environmental licences and consents, which could include, for example, the following.
  - Waste carrier licences.
  - Environmental permits for waste transfer stations.
  - Hazardous waste site registration.
  - Environmental permits for the operation of mobile crushing equipment.
  - Environmental permits for the operation of mobile soil treatment equipment.
  - Environmental permit exemptions for the reuse of construction waste.
  - Environmental permits for discharge.
- ☑ Waste transfer documentation.
- ☑ Environmental inductions, briefings and toolbox talks.
- ☑ Environmental consultations, including meetings and formal correspondence with regulators.
- ☑ Environmental incident reports and non-conformance reports and breaches.
- ☑ Environmental work instructions and operational procedures.
- ☑ Environmental monitoring and inspection records, which could include, for example, the following.
  - Weekly site inspection records.
  - Audit reports.
  - Water sampling records.
  - Soil and waste sampling records.
  - Dust and air quality monitoring, including visual inspections.
  - Visual records of sensitive area protection, including fencing.
  - Records associated with the achievement of project environmental objectives and key performance indicators.

A number of the above documents will need to be retained for legal purposes, such as waste transfer documentation, which must be retained for non-hazardous waste for two years, and consignment notes for hazardous waste, which must be retained for three years. These requirements must be made clear following completion of the project to ensure all relevant documents are archived, together with the project contractual documentation.

Appendix A – Examples of environmental objectives

Objective	Action	Key performance indicator	Performance*			
			Q1	Q2	Q3	Q4
Water Monitor site water usage and aim for a reduction in usage against previous year baseline Offices to identify their usage and identify ways to reduce water demand	■ Sites to continue measuring and reporting on site water usage. (All sites to fit flow meters or utilise existing meters) ■ Projects to identify and implement water saving measures on site ■ Advertise and raise awareness of water efficiency through newsletters and communications ■ On design and build projects, teams are to identify and communicate stakeholder risks and opportunities presented by water use	All projects to report on water usage				
		Projects to identify and implement water saving efficiencies to reduce company usage of water by 5% against previous year's consumption figures/£100k turnover				
		All offices to identify and install water-saving efficiencies				
		Communicate risks and opportunities presented by water to identified clients				
Energy and carbon reduction Monitor energy usage on sites and offices to reduce company energy usage by 5% against previous year baseline	■ Install sub-meters in all offices to record energy usage ■ Promote energy and carbon awareness through training and promotional events ■ All offices to prepare a transport and logistics plan ■ All offices to create a green travel plan ■ Each office to measure the carbon footprint of projects where this has been requested by the client ■ Identify suitable sites for carbon footprint assessment	All new projects and offices to have meters installed (kWh/100 m^2)				
		Each office to create a green travel plan				
		Each office to receive an energy audit and implement energy saving measures				
		Measure the carbon footprint of projects where this has been requested by the client				
Responsible sourcing Monitor the use of responsibly sourced materials	■ Projects to procure timber in accordance with Government's CPET requirements where possible ■ Projects to identify and implement the use of recycled aggregates/ materials, where possible	All projects to report on the percentage (by value) of all site timber having chain of custody				
		All projects to report on volume of recycled aggregates				

*** Key:** 1 Completed/objective met 2 Started 3 Not started

continued

Examples of environmental objectives *continued*

Objective	Action	Key performance indicator	Performance*			
			Q1	Q2	Q3	Q4
Waste Reduce waste production in line with the waste hierarchy and raise waste awareness through training	▪ Set waste reduction targets based on previous year's data ▪ Identify and develop good practice for waste reduction ▪ Undertake supply chain waste reduction workshops ▪ All sites to appoint a waste champion ▪ Project staff to receive SWMP refresher training.	Company to divert at least 85% of construction waste from landfill				
		Sites to reduce waste produced on site by 5% compared to previous year's levels				
Biodiversity Improve company knowledge on biodiversity issues and benefits whilst improving the biodiversity on our sites	▪ Deliver appropriate biodiversity training to teams ▪ Benchmark and improve the biodiversity on projects where relevant (such as BREEAM, client) ▪ Share good practice examples	Deliver relevant awareness training (identified on a project, risk basis)				
		Projects to identify ways to improve the biodiversity on site				
Environmental incident rate Ensure all site personnel are aware of environmental incident reporting requirements	▪ Ensure all sites record environmental incidents, near misses and complaints ▪ Encourage dialogue on environmental issues/ situations that are not easy to implement, are misunderstood or ignored ▪ Include environmental incident/near miss reporting within site inductions and toolbox talks	Target a continued zero incident rate				
		All risk assessments and method statements to include specific environmental risks				
		Toolbox talks – environmental topic ratio 1:4				
Community engagement Increase communication with local communities and maintain a considerate constructor approach	▪ Capture and collate in an annual report all the community engagement work the company and projects carry out ▪ Create a company community engagement plan ▪ Improve compliance with Considerate Constructors' scheme (CCS) by creating a company minimum standard for sites, share good practice and have a dedicated CCS folder for each registered site	Record any community engagement work and send to the environmental team/advisor or head office				
		Each project to have a CCS folder and a target score of 34				

*** Key:** 1 Completed/objective met 2 Started 3 Not started

For an interactive PDF version of this form visit the companion website.

E 10

Appendix B – Site environmental management systems checklist

Company name		Project title	
Location		Contract no.	

Site environmental management systems	Yes	No	N/A
1. Is the company (or site) environmental policy available and displayed on site, with relevant requirements highlighted at the site induction?			
2. Has the contract and client documentation been reviewed to identify the site-specific environmental requirements, including planning conditions?			
3. Has a construction environmental management plan been prepared to identify the relevant objectives, issues and controls?			
4. Is there a system for ensuring that suitable and sufficient arrangements are in place for the management, storage and disposal of waste (such as the production of a site waste management plan)?			
5. Is there a system for ensuring that all site personnel (including sub-contractors) are made aware of the project environmental issues and standards through site inductions, briefings and toolbox talks?			
6. Have responsibilities for environmental management been defined, allocated and communicated through inductions, noticeboards, and so on?			
7. Is someone responsible for ensuring that environmentally sensitive areas are identified and protected (such as drains, rivers, streams, groundwater and areas containing protected plants and animals)?			
8. Is there a system for ensuring that the necessary environmental licences have been identified and granted (such as discharges of water and effluent, tree preservation orders, premises producing hazardous waste and waste management licences)?			
9. Is there a system for ensuring that method statements include sufficient control measures for environmental management?			
10. Is there a system for sites to deal with environmental emergencies (such as for spillages and groundwater contamination) and for notifying them to the relevant authority?			
11. Is there a suitable and sufficient monitoring regime in place to ensure that the site environmental requirements are being met (such as relevant environmental inspections for oil and waste storage, protected areas, water quality, noise and dust)?			
12. Are environmental records being maintained to facilitate good auditing processes (such as waste transfer notes, consignment notes, noise and water monitoring)?			
13. Have adequate arrangements been established for liaison and communication with local interested parties (residents, shops, businesses, schools and parish council)?			
14. Is there a process in place to review ongoing performance and identify areas for improvement?			
15. Does the environmental management system (EMS) extend beyond the site activity to the supply chain for services and materials?			

Comments							
Name		Position		Signature		Date	

(This checklist is based on the CITB Construction site health, safety and environment auditing system (SA03 CD).)

For an interactive PDF version of this form visit the companion website.

E 10

Index